THE HOME UNIVERSITY LIBRARY
OF MODERN KNOWLEDGE

179

SCIENCE IN ANTIQUITY

Science in Antiquity

BENJAMIN FARRINGTON

PROFESSOR OF CLASSICS AT THE
UNIVERSITY COLLEGE OF
SWANSEA

Geoffrey Cumberlege

OXFORD UNIVERSITY PRESS

LONDON   NEW YORK   TORONTO

First published in 1936 and reprinted in 1947

PRINTED IN GREAT BRITAIN

CONTENTS

PREFACE TO SECOND IMPRESSION

The reprint of this book gives me the opportunity to make a few improvements. For the detection of certain errors I am indebted to Professor Lancelot Hogben ; Dr. Herbert Chatley ; and Mr. J. Duff, Sidney, British Columbia, Canada.

B. FARRINGTON

1946

The great intellectual division of mankind is not along geographical or racial lines, but between those who understand and practice the experimental method and those who do not understand and do not practice it.

DR. GEORGE SARTON

CHAPTER I

EGYPT AND MESOPOTAMIA

It has been the almost universal custom until quite recent times to deny the existence of any science before the Greeks. This is no longer possible. It is true that our science can be derived by a continuous tradition from a brilliant flowering in the sixth century B.C. in the Greek town of Miletus on the coast of Anatolia. But it is equally true that before the middle of the second millennium B.C., that is to say, more than 1,000 years before the dawn of Greek science, definitely scientific elements are to be found in at least two of the older eastern civilizations, those of Egypt and Mesopotamia. And though the bridge connecting them with the beginnings of Greek science has been partially broken down through the loss of the historical tradition, that there was a connection, an influence of the older civilizations on the Greek, is certain. Thus, though the main purpose of this volume is to trace the history of Greek science from its origin in the sixth century B.C. till its virtual disappearance about the fifth century of our era, it would be to flout one of the most fascinating results of modern research if one failed to put the Greek achievement in its proper historical setting, to wit, the older civilizations of the Near East.

How the knowledge of this dependence of Greek civilization on the older civilizations of Egypt and Babylon came to be lost is an important question to which an answer will be attempted in our last chapter. But the connection of modern science with the Greek was never lost sight of. For when science came into its own in the seventeenth century it did so under the direct inspiration of the Greeks. Copernicus, Galilei, and Newton were students of Ptolemy, Aristarchus, and Euclid. And Vesalius, the founder of modern anatomy, was loud in proclaiming his debt to the example and instruction of his forerunners among the ancient Greeks. Historians of science, therefore, could not fail to acknowledge the debt of modern science to the Greeks. But with regard to the science of the Egyptians and Babylonians ignorance reigned. The ancient Greeks, indeed, acknowledged a heavy debt for the elements of their mathematical and astronomical knowledge to the civilizations of the Nile and the Euphrates. But modern historians, lacking the means to confirm this tradition, were content to ignore it, or, out of mistaken partisanship for Greece, strained every nerve to deny it. Greek history has often been written as if the mission of Greece had been to save Europe from some dread abstraction called Orientalism. But this is to prolong down the ages a sentiment appropriate only to the generation of Marathon and Salamis. A soberer judgment will recognize the justice of the Greek view, that Greek civilization was a continuation of the older civilizations of Mesopotamia and Egypt.

The means whereby to estimate the extent of the debt of Greece to the Near East has only begun to be put into our hands in recent times by the progress of archæology. First came the recovery of written records of the ancient civilizations on stone monument, baked clay tablet, or papyrus roll. Next followed the solution of the difficult problem of reading them. Assyrian and Persian cuneiform, and the hieroglyphic and hieratic scripts of Egypt have yielded up their secrets, and the material thus recovered, deficient though it still is, has revolutionized our knowledge of the past. When Queen Victoria came to the throne the year 4004 B.C. was generally accepted for the creation of the world. Writing was supposed to have been invented by the Phœnicians about the seventh century B.C. We now possess written documents that go back almost to the date at which men of the Victorian Age supposed the world to have been created.

Thus was restored to us a better knowledge of the history and culture of the old civilizations of Egypt and Chaldea. Their achievement in certain respects was so great and obvious as to be admitted at once without possibility of cavil. That in the third millennium B.C. men could administer large populations, build great cities, and create wonderful works of art, was universally admitted. That literatures of some scope and variety began also to appear in this remote epoch could not long be denied. About 2800 in Egypt lived the philosopher Kegemmi whose book of maxims was a classic ; and the great code of laws of Hammurabi, king of Babylonia, was committed to writing about 2000 B.C. But that anything worthy of the name of science existed before the Greeks was a question about which it was still legitimate to argue until the publication of the evidence in the twentieth century put the matter beyond doubt.

Before briefly discussing the documents of Egyptian and Babylonian science there is a preliminary question about which a word or two must be said. The connection between science and technique is close and important. It is true that the goal of technique, as practical, may be distinguished from the goal of science which is theoretical. The technician wants to do something, the scientist wants to know. But we have come to realize that the best proof that our knowledge is genuine is that it enables us to do something. Science is continually tested in action. We have also come to feel that in its origin science is not in fact so divorced from practical ends as histories have sometimes made out. Text-books, right down from Greek times, have tended to obscure the empirical element in the growth of knowledge by their ambition of presenting their subjects in a logical orderly development. This is, perhaps, the best method of exposition ; the mistake is to confuse it with a record of the genesis of theory. Behind Euclid's definition of a straight line as " one that lies evenly between the points on it " one divines the mason with his level. And the recent discovery of a fragment of the *Method* of Archimedes shows that weighing of solids of different shapes against one another suggested the relationships of volume that were afterwards demonstrated with rigorous logic. It was a practical problem in each case that stimulated the same Archimedes to the invention of his screw, of his system of pulleys, of the idea of specific gravity. Techniques are a fertile seed-bed of science, and progress from pure empiricism to scientific empiricism is so gradual as to be imperceptible.

From this point of view the astonishing achievement in technique of the ancient civilizations must be regarded as a step in the attainment of science. The Egyptians, for instance, had discovered metals as far back as 4000 B.C. Before 3000 B.C. they had an alphabet, and pens, ink, and paper. The age of the construction of the great pyramids is from about 3000 to 2500 B.C. In this epoch the Egyptians were also busy with agriculture, dairying, pottery, glass-making, weaving, ship-building, and carpentry of every sort. This technical activity rested upon a basis of empirical knowledge. For instance, in the third millennium they had various instruments of bronze. These instruments show a constant proportion of about twelve per cent. of tin, which gives the alloy a maximum of hardness without fragility. The fixing of such a proportion is certainly the result of rational observation. To deny it the name of science because it was, perhaps, handed down by tradition to apprentices instead of being written in a book is not wholly just. Even in the sixteenth century of our era Vesalius was so impressed by the value of practical apprenticeship over instruction out of books that he felt compelled to apologize for printing in his Anatomy the wonderful plates which are among the chief glories of modern science. He only did so, he says, because boys were no longer apprenticed to doctors and trained by direct operation on the human body.

Not only had the Egyptians a knowledge of the right proportions of copper and tin to secure the most serviceable bronze, they had also a technique of tempering the alloy by various processes designed to secure toughness, hardness, or flexibility. Technical problems also certainly clamoured for solution in connection with their gold-work, weaving, pottery, hunting, fishing, navigation, basket-work, culture of cereals, culture of flax, baking and brewing, vine-growing and wine-making, stone-cutting and stone-polishing, carpentry, joinery, boat-building, and the many other processes so accurately figured on the walls of the tombs of the nobles at Sakkara (2680 to 2540 B.C.). In all these techniques lay the germ of science, as also in the accumulated experience which enabled them to rear the huge pyramids, excise and transport gigantic monoliths, and invent the bellows, the siphon, and the humble shadoof or well-sweep.

But all this technical knowledge, important as it is, is still not science in the full sense. It contains no hint of an attempt to explain all the phenomena of the universe on the basis of an intelligible system of natural law, which is the goal of positive science. And it may be remarked in passing that we have no proof of anybody having risen to this conception before the Greeks in the sixth century B.C. But, apart from this, we have as yet no proof, in all this evidence from technique, of the attempt to organize even a particular branch of knowledge in a scientific way. The technical achievement in itself is not proof of the power of conscious abstraction, of the capacity to detect general laws underlying the variety of phenomena and to utilize these general conceptions for the organization of knowledge. To put the point in another way, we have no evidence derived from the various techniques that have been mentioned that the Egyptians were attempting to classify the various substances of which they were aware and describe their properties, or to do the same for plants or animals. We have no evidence that they were asking how one thing could apparently change into another, how bread, for instance, which a man ate could turn into flesh and blood. We have no evidence that the siphon prompted them to consider the possibility of a vacuum or to rise to such a generalization as that Nature abhors a vacuum. In short, so far as the evidence from techniques goes, we have no certain proof that the Egyptians possessed that kind of curiosity and that gift for speculation which are necessary for the creation of science in the full sense. Interesting and important as their techniques are, essential as they are for creating that wealth of particular knowledge without which science cannot be, they do not constitute truly scientific knowledge, and without additional evidence we should be justified in denying to the Egyptians the credit of having stepped over the threshold of the temple of science.

This additional evidence is to some extent supplied by the achievement of the Egyptians in astronomy. A highly organized state like that of the Egyptian Pharaohs is impossible without a calendar, and the introduction of the first practical calendar is now authoritatively placed at 4236 B.C. The seasonal activities of the Egyptians were necessarily dependent on the rising of the Nile. This roughly coincides with the summer solstice, and the solstice itself coincides with the heliacal rising of the bright star Sirius. That is to say that Sirius, after a period of absence from the night sky, is seen for the first time just before sunrise above the eastern horizon at the summer solstice ; and this heliacal rising of Sirius announces the rising of the Nile. This double coincidence was early observed by the Egyptians and used by them for a neat method of correcting the slight inaccuracies of their calendar. The Egyptian year was divided into 365 days, or twelve months of 30 days, plus 5 extra " heavenly " or sacred days. This calendar year was employed by the Egyptians for co-ordinating the many activities of their wealthy and populous realm. No attempt was made, as none is made with us, to make the days of the month correspond with the phases of the moon. Neither was any attempt made to make the year of 365 days correspond with the solar year, which exceeds it by a quarter of a day. We make the necessary adjustment by the insertion of an extra day every fourth year. The Egyptians simply kept a record of the divergence which adjusts itself automatically every 1,460 years (365 × 4 = 1,460). The divergence was noted by direct observation of the heliacal rising of Sirius which every fourth year would occur a day later by the calendar. This great cycle of 1,460 years, the interval after which the solar year and the calendar again coincide, is known as the Sothiac period, from Sothis, the Egyptian name for Sirius. It is certainly evidence both of astronomical knowledge and practical ingenuity on the part of the Egyptians. But since the Egyptian astronomy was almost certainly based on Babylonian observations we shall not lay too much stress on it as proof of their scientific capacity.

The unmistakably scientific achievement of the Egyptians lies in the domains of mathematics and medicine. Our knowledge of it rests on the chance preservation of a few papyrus rolls, or fragments of rolls, of which the most important for mathematics is that known as the Rhind papyrus, while the only one of importance for medicine is the Edwin Smith papyrus. These two documents were found together in the middle of the nineteenth century, so that our whole knowledge of Egyptian written science depends on one discovery. This reminds us in a forcible way of the accidental character of our information, and warns us to keep an open mind as to the possible extent of Egyptian science.

The mathematical papyrus was written probably about 1650 B.C. but copied from an original earlier than 1800 B.C. Its interpretation gives rise to considerable difficulty, but it is abundantly evident from it that the Egyptians had an elementary science of mathematics. They had a decimal system of notation with signs for 1, 10, 100, 1,000, 10,000, 100,000. The number of units, or tens, or hundreds, etc., was expressed, as in Roman numerals, by repeating the sign the desired number of times. The operations of multiplication and division were performed by a series of additions and subtractions. They could handle fractions though with difficulty, all proper fractions being reduced to groups of fractions with 1 as numerator. Thus $\frac{2}{29}$ was expressed as $\frac{1}{24} + \frac{1}{58} + \frac{1}{174} + \frac{1}{232}$

They had also some knowledge of the properties of numbers. They knew that the squares of 3 and 4 add up to the square of 5, and by the method of proportion they were able to make use of this knowledge in calculations. They connected this knowledge with the right-angled triangle, realizing that in a triangle with sides 3, 4, and 5, the side of 5 units subtends a right angle. They were thus familiar with a particular case of the theorem of Pythagoras, and they used this knowledge to enable them to erect perpendiculars.

In the Rhind papyrus certain problems of arithmetic, geometry, and mensuration are solved. One of them seems to show an application of mathematics to an ancient technique, that of pyramid building, which dates from the beginning of the third millenium. Supposing one had to face a pyramid with stone, at what angle should the stone be cut?

It is a problem in similar triangles. If the base of the pyramid is 10 units, and the perpendicular height 8 units, the problem can be solved, as in the accompanying illustration, by marking off 5 units on the base of the stone to be cut, erecting a perpendicular of 8 units, and completing the triangle.

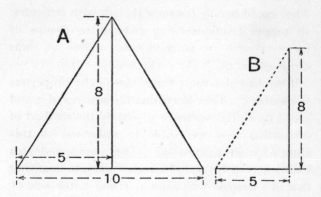

A gives the proportions of the pyramid. B shows the construction for finding the angle at which to cut the stone.

The glimpse into Egyptian medicine provided by the Edwin Smith papyrus is even more surprising than their mathematics. The papyrus is a fragment of a surgical treatise dealing with injuries to the body which are classified by the simple and natural process of passing from the head downwards. It deals with the skull, the nose, the jaw, the ears, the lips, the cervical vertebræ, the collar-bone, shoulder and shoulder-blade, the thorax and breast ; and as the discussion of the vertebral column begins the MS. breaks off. Each type of case is dealt with systematically. The injury is mentioned, the symptoms which examination should reveal are described, next comes the diagnosis, and then the verdict as to whether the complaint is curable, possibly curable, or incurable. Finally the treatment is set forth.

The anatomical knowledge is correct and considerable in amount. There is a beginning of physiological knowledge, the most remarkable sentences being those in which the body is conceived of as being a single organism with the heart as its central organ. The therapeutical means are meagre but totally free from magic. The whole treatise is purely positivistic in spirit. The mind of the writer is subdued to the acceptance of the observed fact, as is particularly noticeable in the recognition of certain states as incurable. We have here a body of knowledge which can only be regarded as the result of a long tradition of observation and reflection. As such it is a work of science in the modern sense. It should be added that the document bears in itself the proof that its teaching is not new. Technical terms are interpreted for the reader. It is therefore presumably a manual dealing with a traditional branch of knowledge which may quite possibly be as old as the fourth millennium B.C.

Babylonian science, to which we must now turn, is possibly older than Egyptian, and its records are more abundant. The Egyptians wrote with pen and ink on reed-paper, a perishable material. In the kingdoms of Sumer, Akkad, and Assur, the successive centres of what is for convenience called Babylonian culture, writing was done with a stilus on soft clay tablets which were then baked and formed practically indestructible documents. The remains of Assurbanipal's library (died 626 B.C.), now in the British Museum, number 22,000 clay tablets. At the temple library of Nippur some 50,000 tablets were found ranging in date from about 3000 B.C. to 450 B.C. On such material our knowledge of Babylonian science is based.

The Babylonians like the Egyptians were acquainted with a decimal system of notation. They supplemented this with a sexagesimal system which is among the most astonishing inventions of ancient science. Their decimal notation, which was essentially the same as that employed by the Egyptians and, later, by the Greeks and Romans, was used for numbers from 1 to 59. For numbers greater than 59 they employed their sexagesimal notation of which the bases are 1, 60, 3,600 (60^2), 216,000 (60^3), etc. The numbers from 1 to 59, as in the Egyptian, Greek, and Roman systems, might be written in any order ; in the sexagesimal notation the value of the symbols, as in our modern notation, depends on the order. Symbols in the first position are units, in the second position units multiplied by 60, in the third units multiplied by 60^2. Thus the cube of 16, i.e. 4,096, is written 1, 8, $16 = 1 \times 60^2 + 8 \times 60 + 16 = 4,096$. For the positional notation a zero sign is required; the Babylonians had an equivalent for this.

In its combination of the decimal and sexagesimal systems Babylonian notation is clumsy. But that the use of position in writing numbers should have been known in all probability in the third millennium in Mesopotamia is a very remarkable circumstance. The Greeks, the greatest mathematicians of antiquity, never hit upon this device, and it was introduced into Europe only about the eighth century A.D. from the Hindus. The choice of so large a base as 60 is also surprising. If, as seems likely, the Babylonians were influenced in this choice by the fact that 60 is divisible by 2, 3, 4, 5, this is additional evidence of the conscious ingenuity that presided over the elaboration of their unique notation.

A characteristic of Babylonian mathematics is the fondness for tables, like our multiplication table. We find among their remains tables for multiplication and division, tables of the squares of all the integers up to 60, a table of cubes of the integers up to 16, not to speak of tables of square roots and even a table of cube roots. There are also examples of arithmetical and geometrical progressions. We are therefore in presence of abundant results of their mathematical ability, but the tables are offered to us, like our own practical tables for calculating interest and so forth, without proof or theory, so that, as far as the evidence goes, Babylonian arithmetic lies under the suspicion of being largely empirical.

The impression of Babylonian mathematics as a science emerging from an empirical stage is confirmed by their geometry. We have abundant proof of the practical capacity of the Babylonians in mensuration. They could measure accurately fields, or building lots, of irregular shape. Their method was to divide the area into right-angled triangles, rectangles, and trapezes, the areas of which they knew how to determine. But on two tablets which date from about 2000 B.C. we find something more purely theoretical. These tablets give formulæ for determining the length of the hypotenuse of a right-angled triangle in terms of the other two sides. This amounts to nothing less than an attempt to solve the famous theorem of Pythagoras. The Babylonian method instructs us in the tentative process by which the human mind advances to the idea of a general solution of such a problem. The tablets offer us two formulæ, both empirical, both yielding only approximately accurate results, and both attempting to solve only a particular triangle, namely that in which the sides that contain the right angle are in the proportion of 10 to 40. We are thus still very far from the famous solution of prop. 47, Euclid Bk. I, but a long step has been made in its direction. The Babylonians have not been content to measure the length of the hypotenuse directly, which would be pure empiricism. They have attempted to determine it *a priori*, to establish a rule, to provide a formula which would save the bother of measuring for ever. They have realized clearly that there is a permanent relationship between the lengths of the sides of a particular type of triangle. In science the first steps are the most difficult, and the statement of a problem is a contribution to its solution.

To the Babylonians also is due the division of the circumference of the circle, and of the four right angles subtended by the circumference at the centre, into 360 parts or degrees. This capital invention was taken over by the Greeks and passed on by them to western Europe. If we couple with this invention the fact that almost all Babylonian measures, whether of length, surface, volume, capacity, or weight, are based on 3 or 12, we shall be the more impressed with the extent to which they applied number to the regulation of their complex material civilization. The uniformity here observable fully entitles their convention to be called a *system* of weights and measures. And this systematization, with all that it implies of comparison between lengths, surfaces, volumes, angles, and weights, is an enormous step along the path to modern science. It implies a conception of the abstract quality of number, as well as a wide-spread popular education in the manifold uses to which number can be put.

The extent to which Babylonian science had been penetrated by the importance of number and exact calculation is nowhere more strikingly exhibited than in their astronomy. Their astronomy certainly had its superstitious side. It was developed in the interest of the now rejected ideas of astrology. But the extravagances of astrology have a justification which must not be overlooked. The phenomena of the heavens have an obvious influence on the earth. The alternations of the seasons and of light and dark are dependent on the sun, so that all life, animal and vegetable, is visibly subject to the great source of heat and light. The moon is the measurer of time, and where the phenomena of tides are observable they are apt to be connected with its influence, as is also the physiological cycle of women's life. To seek to extend this influence to the planets is natural, and when we call a man jovial or say that he has a mercurial temperament we are using the language of this ancient belief. The Babylonian priesthood of the Chaldeans were the great practitioners of astrology in antiquity, and many of their practices came rightly to be condemned as superstitious ; but from their age-long observation of the skies sprang a true science, which independently of its later Greek developments is of the greatest importance in the story of man's intellectual conquest of his world.

Among the tablets found at Nineveh, which go back to the time of Sargon I, about 2800 B.C., are the fragments of a veritable treatise on this ancient science. Here we find rudimentary calendars of the aspects and risings of the various constellations, from which predictions relative to the king and the country were to be drawn. The planets also were supposed to be of special importance for human life. Accordingly the motions of the planets were carefully observed. The planets as they revolve round the sun (a fact of course unknown to the Babylonians) necessarily present to us the appearance of following the sun's annual path at varying speeds, of halting, and even of reversing their courses. These irregularities in the motions of the planets, their retrograde motions, and their halts, were all observed and recorded before the twelfth century B.C.

The Babylonians had a luni-solar year of 12 months of 30 days each. This gives but 360 days to the calendar year ; accordingly every six years a thirteenth month was intercalated to keep the calendar in accord with the seasons. This problem of the regulation of the calendar, to which we shall recur in later chapters, led the Babylonians to ever more accurate determinations of the length of the month and of the year. In addition to the necessity imposed by the calendar of making precise observations of the courses of the sun and moon, the Babylonians had also astrological reasons for studying and recording the behaviour of these luminaries. Eclipses of the sun and moon were supposed to be full of import for the destiny of the royal family and the country; and it was above all important not to be taken unawares by the occurrence of an eclipse. Accordingly the cycles in which the eclipses of sun and moon recur were early determined with fair accuracy. It is to the Babylonians also we owe the knowledge of the ecliptic, or the apparent yearly path of the sun through the fixed stars. To help them in their determination of the ecliptic they mapped the stars that lay on either side of it into the twelve signs of the Zodiac, the breadth of the Zodiacal belt being determined by the distance to which the planets appear to move on either side from the path of the sun.

These capital inventions of the ecliptic and the Zodiacal belt were given precise mathematical usefulness by the Babylonian device of dividing the circumference of the circle into 360 degrees. The angular distances of the heavenly bodies from one another could thus be observed and plotted with precise accuracy. A most impressive application of mathematics to astronomy is supplied by a tablet found in the library of Assurbanipal at Nineveh. The library belongs to the middle of the seventh century, but the document may be itself much older or a copy of an older document. It is an attempt to tabulate the progress of the illumination of the surface of the moon during its period of waxing. To this end the area of the moon's face is divided into 240 parts, over which the illumination is conceived as spreading first according to a geometrical, then according to an arithmetical progression. This arrangement does not correspond to the facts. Without the possession of far more accurate instruments than were at the disposal of the Chaldean observers it could hardly do so. Nevertheless this early endeavour to *measure* a physical event is of the utmost importance in the history of science.

It may be asked what instruments the Babylonian astronomers used for the creation of their remarkable collection of observations. They had three instruments, which seem to be of their own invention, the sun-dial, the clepsydra or water-clock, and the polos. The last is a refinement of the sun-dial. It consists essentially of a hemispherical bowl with a vertical pointer at the bottom reaching exactly to the centre of the sphere. The shadow of the tip of the pointer traces out on the bowl in an inverse direction the course of the sun through the heaven. Further, the polos was fitted with an adjustable armillary sphere, that is to say, an open-work or skeleton sphere, representing the Zodiacal belt with its twelve signs and its division into 360 degrees. This could be used as a clock by night. All that was necessary was to know the sign and the degree in it occupied by the sun at the moment of setting. At any hour during the night the armillary sphere could be adjusted so that the signs on it occupied the same position as the signs observable in the night sky. As it was moved round to secure this adjustment, the degree occupied by the sun at its setting would move across the hour lines marked on the polos, just as the shadow of the pointer did by day. In this way the time could be read from the stars at night as it was read from the sun by day.

The motive behind the Babylonian study of the skies was, as we now see, partly mistaken ; but their observations, extending over hundreds, even thousands, of years, were recorded with accuracy and became the material of true science. The supposed connection between eclipses of the sun and moon and the motions of the planets and the fortunes of men on earth has not been borne out by experience, although there are still educated people who have their horoscopes cast. But a true science of the connection of one celestial event with another celestial event gradually emerged. This was the first demonstration of a mathematical regularity in the ordering of the phenomena of the universe, and it gave man his first great lesson in the method and the possibility of science.

St. Augustine, in his *De Doctrina Christiana*, while warning his readers against the superstitious practices of the astrologers, recognizes and well describes the truly scientific element in the old Chaldean lore. " A knowledge of the stars," he says, " has a justification like that of history, in that from the present position and motion of the stars we can go back with certainty over their courses in the past. It enables us with equal certainty to look into the future, not with doubtful omens but on the basis of certain calculation, not to learn our own future, which is the crazy superstition of pagans, but so far as concerns the stars themselves. For just as one who observes the phases of the moon in its course when he has determined its size to-day, can tell you also its phase at a particular date in past years or in years to come, so with regard to every one of the stars those who observe them with knowledge can give equally certain answers." Under the patronage of St. Augustine astronomy, at least in so far as it concerns the apparent positions of the heavenly bodies, entered into the curriculum of Christian education.

We have now spoken briefly of the scientific achievement of the Near East. Of what scientific elements there may have been in the ancient civilizations of China and India it is hardly possible to speak in a volume the size of this. The study is as yet insufficiently advanced, and there are in particular difficulties of dating that remain unsolved. What does seem clear is that some three or four centuries before the beginnings of Greek geometry there was current, not only in Egypt and Babylonia, but also in India and China, some apprehension of the relationships expounded in the theorem of Pythagoras. The degree of this knowledge, and the possibility of its diffusion from a common centre, are questions that may one day be answered with a confidence that is now impossible. But when the answer is given, if it ever is, perhaps neither Babylon nor Egypt will appear as the earliest exponent of civilization. The Nile and the Euphrates may have to yield place to the Indus. In recent years excavations in the valley of this great river have brought to light the ruins of two cities, Mohenjo-Daro and Harappa, lying 400 miles apart, which proclaim the existence of a hitherto unsuspected civilization dating at least from the third millennium, which, it is claimed, rivals those of Egypt and Babylon. When the monuments of this civilization, including its writing, have been explored and interpreted, the record of scientific achievement before the Greeks may need to be written again.

CHAPTER II

EARLY GREEK SCIENCE

PART I : THE IONIAN SCHOOLS

IT has become almost a habit in discussing the beginnings of Greek science to speak of " the Greek Miracle." The phrase is an unfortunate one with which to describe the great intellectual advance which then took place, a characteristic of which was that it banished miracle from nature and history and substituted law in its place. Even as a means of emphasizing the originality of the Greek achievement the word is too strong. The " miracle " was well prepared for by the Egyptians and Babylonians. Already in 1883 A. H. Sayce (in *The Ancient Empires of the East*) was contending for the recognition of what was the tradition of the Greeks themselves, " that Greek history and civilization are but a continuation of the history and civilization of the ancient East." And, as we have seen, fifty more years of archæological research have fully borne out his contention. Even with regard to the special subject of this study the Greeks cannot be looked upon as the earliest pioneers. Something both of the method and results of science

C

was included in the tradition of which they were the heirs.

Those who still write the history of science as if it began among the Greeks often open their story by referring to the elements of scientific knowledge that may be gleaned from the pages of the earliest monuments of Greek literature, the *Iliad* and the *Odyssey*. There we find the knowledge of a few stars, and mention of medicine and surgery and of professional physicians. Anatomical knowledge is revealed by the existence of some 150 special terms for different parts of the body, mentioned as injured in war. We hear also of the smith, the carpenter, the potter, the worker in leather ; of spinning and weaving ; of the use of animal manure in agriculture ; of gold, silver, lead, iron, steel, bronze, tin, and amber. But it is obvious that there is nothing in all this of much importance for the history of science, nothing that cannot be matched at a much earlier date among the Egyptians and Babylonians.

Nevertheless with the Greeks a new and most important element did enter science. This is the element of speculative philosophy, which constitutes the specific quality, the real originality, of Greek science ; and this, in the opinion of the present writer, is most definitely to be associated with the *Iliad*, which embodies a profoundly original world-view that runs like a leaven through Greek thought.

The great originality of the *Iliad* lies in this, that the events of which the story consists are represented as springing out of the characters of the actors. In the conception of the author historical event and human character form a complex, a fusion, in which the elements can be analysed but not separated. This conception makes man in some sense the author of his own destiny and no longer a puppet in the hands of Fate. It therefore demands of the poet that he be a dramatic poet, a creator of character, and it tends to heighten the self-consciousness of the reader. Nothing could be more opposed to the fatalism of Chaldean astrology than the conception of human will and character that dominates the *Iliad*. The note is already clearly sounded that Shakespeare echoes in *Julius Cæsar* :

> Men at some time are masters of their fates :
> The fault, dear Brutus, is not in our stars,
> But in ourselves, that we are underlings.

The Achilles of Homer is a man who has chosen his fate. Better a short life with honour than a long and inglorious one. The chivalrous Hector makes his choice in the course of the poem. And the heroic temper of Ajax is revealed, as Greek criticism pointed out, in his prayer to Zeus when a thick mist has enveloped the field of battle : " Make clear day, and vouchsafe unto us to see. So it be but in light, destroy us ! " So it be but in light.

The cry echoes down through the Greek literature of intellectual emancipation in a hundred formulæ which proclaim knowledge the basis of the good life. It is the burden of the best sentence Plato ever wrote : " An unexamined life is no life for a man."

This conception of man as in some measure master of his fate gives to the human characters in the *Iliad* a moral grandeur that is new. Beside them the figures of the Homeric gods fade into insignificance. They are felt rather as poetic machinery than as objects of religious worship or superstitious awe. This secular note in the *Iliad* is very pronounced, and it is also characteristic of Greek scientific thought in its beginning in the sixth century B.C. The organized knowledge of Egypt and Babylon had been a tradition handed down from generation to generation by priestly colleges. But the scientific movement which began in the sixth century among the Greeks was entirely a lay movement. It was the creation and the property, not of priests who claimed to represent the gods, but of men whose only claim to be listened to lay in their appeal to the common reason in mankind. The Greek thinker who advanced an opinion stood behind the opinion himself. He claimed objective validity for his statements ; but they were his own personal contribution to knowledge and he was prepared to defend them as such. Consequently with the Greeks individual scientists begin to emerge, and the specific quality of scientific thinking begins to be recognized.

To put the matter in another way, the worldview of the Egyptians and Babylonians was conditioned by the teaching of sacred books ; it thus constituted an orthodoxy, the maintenance of which was in the charge of colleges of priests. The Greeks had no sacred books, but as the well-spring of their intellectual life we find a highly individual poem which did, indeed, retain as poetical machinery a view of superhuman powers playing with men as with puppets, but which treated with supreme seriousness the actions and characters of men. This poem was composed, perhaps in the ninth century B.C., in Ionia. It can hardly be an accident that it was in this same country some three centuries later that men using the same idiom made the first effort to explain nature without invoking the aid of supernatural powers. The Greeks themselves felt the connection. A religious poet, Xenophanes of Colophon, writing in the generation after Ionian science had been launched and lamenting the irreverence of the *Iliad*, protests bitterly, " All from the beginning have learned of Homer." Homer created humanism, and humanism created science. Science is essentially an effort of man to help himself. Homer, in the *Iliad*, emancipated man from the tyranny of the gods he had dreamed in the childhood of the race, in teaching him to look upon himself

37

as in some degree the shaper of his own future. For a few centuries man advanced with some confidence along the path of knowledge recognizing the power knowledge gave him. But when the pendulum began to swing back again, when man began again to bow down before the idols of his own making, when not so much the graven images he had made but the books that he had written came to be looked upon as divine, then humanism was at an end and science too.

Homer, in the *Iliad*, provided Ionian science with a background of secularism which was a prerequisite of its coming to be. In the *Theogony*, or *Genealogy of the Gods*, the work of another poet, this time not from Ionia but from the mainland of Greece, we get another feature of Greek thought which was of vital importance in the development of their scientific outlook. According to this poem Greek religious consciousness was without a theory of creation. We are told that in the beginning was Chaos, that Chaos produced Earth, that Earth produced Heaven, and that from their embraces sprang the rest of all the things that are. The idea is that of an evolution or spontaneous development, conceived of in the poem in a mythological way but offering no obstacle to the growth of positive science.

The *Theogony* was probably composed in the eighth century. From the middle of the seventh century on there is an abundant poetry of a new kind, elegiac and lyric, unhappily surviving now only in fragments, which differs from the poetry that preceded it in that the singers now sing of themselves, making of their own actions and passions the subject-matter of a vivid personal poetry of an intimately revealing type. The greatest of this school were Archilochus, Sappho, and Alcæus. Their poetry is of unsurpassable independence of character and maturity of self-knowledge. " For a brief period of splendour," says the Introduction to the *Oxford Book of Greek Verse* " those who had the leisure and the desire for writing had also that perfect understanding of themselves which is the basis of all personal poetry." It was in the age of the great lyric poets—Thales was a contemporary of Sappho—that Greek science began, and there is in it the same boldness of vision and independence of thought. Science is the creation of a particular type of man and of a particular type of society. It does not spring up in a vacuum ; and Greek science will seem less miraculous if we remember the time and place of its origin. It originated in the city of Miletus, on the Greek coastal fringe of Asia Minor. This city was in direct contact with the older civilizations of the east ; it was linguistically a partner in the culture that had already to its credit a brilliant literature of epic and lyric poetry ; and it was a centre of mercantile and colonizing activity. Greek science was thus the product of a rich humanism, a cosmopolitan culture, and an enterprising conduct of affairs.

The names of the three citizens of Miletus who constitute the Early Ionian School, are Thales, Anaximander, and Anaximenes. Their activity covers roughly the first half of the sixth century, and they are traditionally represented as a succession, each being the direct pupil of his predecessor. Whether we choose to regard them as philosophers or scientists is of no moment, for Greek philosophy and science can hardly be distinguished from one another at this date.

Thales is reputed to have made important contributions to geometry. To these we shall refer later. What is of immediate interest is that he is the first man known to history to have offered a general explanation of nature without invoking the aid of any power outside nature. His theory was that everything is ultimately water. He recognized three forms of " that which exists," mist, water, and earth, and he was of opinion that mist and earth are both forms of water. Adapting an idea found in Egyptian tradition, he taught that the universe is a mass of water in which our world forms a bubble, with the earth floating on water at the bottom and the waters above, from which the rain comes, arched over it. The heavenly bodies, which he supposed to be watery exhalations in an incandescent state, float across the waters above just as the earth floats on the waters below ; and sun, moon, and stars, when they set, do not pass under the earth but float round it out of sight till they come to their appointed places on the eastern horizon.

These teachings, so primitive in content, are in spirit very new. Underlying them is a group of simple observations, such as the evaporation and freezing of water, which seems to show the possibility of water existing also as mist and as " earth " ; or the alluvial deposit at the mouth of rivers, which seems to show earth gaining at the expense of water ; or the fact that all living things perish without water, and so are, in some sense, water. Out of such observations he attempts an explanation of all that is. Previous peoples had told stories in explanation of the phenomena of nature, invoking the power of gods to explain events. Such stories had become religious traditions guarded by a professional caste. Thales invokes no power outside nature herself ; his theory is advanced as being his own ; to justify it he appeals to the experience of every man ; nobody is required to accept it unless he finds it true.

Thales had learned some astronomy from the Babylonians and introduced it to the Greeks. But he had gone beyond his teachers in the all-important respect that his teaching was not only concerned with the motions of the stars but attempted to describe their substance. Earth and heaven are for him all of a piece. Like everything else the heavenly bodies are ultimately only water. This is a tremendous enlargement of the domain of science, and constitutes a revolution in outlook. For the Babylonian astrologers the stars were gods ; for Thales they were steam from a pot. Not only were the motions of the heavenly bodies reduced to order, but their substance was made familiar to man. It is a new birth of science.

Profiting by the teaching of Thales, Anaximander advanced in the course of a lifetime of study and reflection to an elaborate cosmology, or theory of the universe, which he embodied in a work entitled *On Nature*. His advance from the position of Thales is along two lines, observation and logical abstraction. He had seen more and thought more.

The step forward in logic is a brilliant one. Instead of explaining all states of matter in terms of one state, as Thales had done, Anaximander derived everything from a primary substance which he called the Unlimited or Indeterminate. This is a great advance towards the abstract conception of matter. The underlying substance is no longer a visible, tangible state of matter but a sort of lowest common denominator of all sensible things arrived at by a process of abstraction. It cannot be wholly understood by being seen or felt, it must also be apprehended by the mind.

Anaximander was of opinion that the Indeter-

minate must be eternal, infinite in amount, and
endowed from the beginning with motion, which he
probably conceived of as motion in a circle. From
these premises he proceeded to give an account of
the evolution of the universe to its present state.
His idea was that the circular motion with which
the Indeterminate is endowed produced a first
stage of " determination " in which the hot was
separated from the cold and formed a sort of envelope about it. This was a deduction from the
observed facts that fire leaps upward and that the
great fires, sun, moon, and stars, lie on the outer
limit of the world. The effect of hot on cold, as
observation teaches, is to cause evaporation. The
sea, he taught, is in continual process of evaporation
and will one day all be dried up. This event, however, lies in the future ; an event that has already
occurred as a result of evaporation is that the
vapours have accumulated and burst the fiery
envelope which enfolded the world. The fiery
envelope has survived as wheels of fire which girdle
the world. But these wheels are not visible to us
as such, because they are wrapped about by tubes
of the moist vapour which burst the fiery envelope.
These opaque tubes are rent in places, and through
these rents the fire gleams down on us. Such is
the real nature of the heavenly bodies. And as the
fiery wheels in their casing of mist revolve about
the earth they produce the phenomenon of the

nightly transit of the stars and of the courses of the sun and moon. Eclipses, and the phases of the moon, are produced by the complete or partial closing of the vents.

It would be possible, even from the fragments that have come down to us, to expand the details of this grandiose scheme, but enough has been said to make its general nature clear. It is an early attempt at a natural history of the universe, essentially of the same kind as the nebular hypothesis of Kant and Laplace. Readers of the first chapters of Wells' *Outline of History* will find an account of the history and structure of the universe much richer, of course, in detail, but precisely similar in outlook.

Another evidence of the advance in logic made by Anaximander is this, that Thales had thought it necessary to support his earth on water, while omitting to tell us on what the water rested ; but Anaximander boldly dispenses with the need of support. His world is poised in space and remains there because of " the similar distance from every-thing," to quote the phrase that has come down to us.

Finally, we have from Anaximander a theory of the origin of life which has startlingly modern features. Living creatures, he taught, came from the moist element when it had been partially evaporated by the sun, and gradually adapted themselves to life on land. That the sea had once covered more of the earth's surface than in his day was deduced by Anaximander from the presence of shells and marine fossils above sea level. He saw a proof for his view that man must have originated from some other animal in the fact that the young of man is unable to fend for itself. The point seems to be that man could not have emerged from the moist element as adult man, because man cannot live except on land ; nor could human babies have been spawned from the sea, for they would have perished. But we have only broken fragments of his thought, of which two further examples may be quoted here. " The first animals were produced in moisture, and were covered with a spiny tegument ; in course of time they reached dry land. When the integument burst they quickly modified their mode of life."—" Living creatures were born from the moist element when it had been evaporated by the sun. Man, in the beginning, resembled another animal, to wit, a fish."

The third of the Milesians, Anaximenes, has not left us anything so impressive as his predecessor. Indeed his logic seems to have been less powerful than Anaximander's. Thus he rejected the theory that the earth hangs freely in space, and he fell away from the conception of the Indeterminate to a view on the level of the water theory of Thales, namely that everything is mist. But he added one brilliant doctrine which redeems his system from the charge of being a mere reaction. According to him it is as the result of a process of *rarefaction* or *condensation* that the primary substance assumes different forms. Mist which is rarefied passes into fire ; if it be condensed it changes first into water and then into earth. Everything depends on the amount of the primary substance that is made to occupy a given space. In this way was the stock of ideas increased by which man might make his universe intelligible to himself.

In one sense the natural philosophers of Miletus inhabited a world which seems very small to us. They had no suspicion of the enormous size of the heavenly bodies relative to our earth, nor of their immense distances from us. The author of the *Theogony* had made the Heaven spring out of the Earth, and the philosophers of Miletus did likewise. The whole heaven was in the nature of an exhalation or evaporation from the earth, and their astronomy was thus hardly distinct from meteorology. But they compensated for this by teaching that there is an infinite number of worlds, which are always coming into being and ceasing to be. Still more important, their method was right. Celestial and terrestrial phenomena were for them essentially the same. One of the fatal wounds to ancient science was the loss of this point of view.

It is of the nature of scientific progress that every new advance raises fresh problems. To one such we now come which is still a major problem for philosophy and science, and which, in the attempts that have been made to solve it, forces the scientist to become a philosopher and the philosopher to become a scientist. The teachers of the Milesian school had advanced very impressive views, but how was one to know that they were true? One could not simply see that they were true, for their theories were in the nature of an argument from things seen. One could not see the fiery wheels of Anaximander; one could not see that the earth was poised in space.

The Milesian philosophers do not seem to have raised this problem at all. They exercised their wits on what their senses offered to them without discussing the question of the validity of their sense knowledge or of their arguments. It was another Ionian philosopher, Heraclitus of Ephesus, who was the first to distinguish clearly between the senses and the reason. He was known as Heraclitus the Dark, and one of his dark sayings is to the following effect : " Of all those whose discourses I have heard there is not one who attains to the understanding that wisdom is apart from other things." If we attempt to interpret this saying in its historical context it seems to mean something like this : Earth exists, water exists, fire exists, mist exists.

We see and feel these things. But we do not see or feel the Indeterminate of Anaximander, nor the principles of Rarefaction and Condensation. These are ways of understanding things, not of perceiving things. If these things constitute wisdom, then wisdom is something different from other things.

Heraclitus does not seem to have been as interested in observation as were the men of Miletus. He was an aristocrat, even of royal blood. Thales was a merchant, and he is reputed to have instructed the Greeks in an improved method, borrowed from the Phœnicians, of taking their bearings at sea. Anaximander made the first map. Heraclitus perhaps despised this manifold activity and busy curiosity. " Much learning," he observed, " does not bring understanding." And it was wisdom, or understanding, in which he specialized. He seems to have derived his own theories mainly from reflection and to have expressed them in riddles. Nevertheless, his cosmological theories bear a general resemblance to those of the Milesians. They could never have existed if Thales and Anaximander had not used their eyes to such good effect.

Heraclitus was of opinion that the primary substance is fire, and that by a process of condensation it passed along the Downward Path to mist, water, and earth, and again by a process of rarefaction along the Upward Path to water, mist, and fire.

So far there is nothing except an adaptation of the theories of his Milesian predecessors. He had, however, profoundly original ideas. In the first place he emphasized strongly the impermanence of things. His teaching on this point is traditionally summed up in the words, which may be his own, " everything is in a state of flux." This theory of the impermanence of things was also knit into his view that the evidence of the senses is deceptive, since it is concerned with impermanent things. Then in the second place he taught that the relative permanence that can be found in things is due to what he called Opposite Tension. Everything, so long as it remains what it is, is balanced between the forces that would take it along the Upward or the Downward Path. This brilliant theory of a comparative permanence maintained in the flux of things by the balance of opposing forces is the dominant thought of his system. A knowledge of this was what he called wisdom, and this wisdom was the gift not of the senses but of the reason. " The eyes and ears are bad witnesses for men," he taught, " if the mind cannot interpret what they say." But the message of the senses, when properly understood, brings men not into a private world of their own but into contact with objective reality. To quote again : " Reason is common (i.e. the same for everybody) ; but men live as if understanding were private to themselves ; now understanding is

D

nothing else but the exposition of the way in which the universe works."

This Heraclitean opposition of the reason and the senses henceforward becomes a major issue in the development of Greek thought. It must not be supposed, however, that it is yet quite the same as our distinction between mind and matter, if by mind is meant an immaterial principle. No Greek had yet got to the point of admitting the existence of any immaterial thing. Heraclitus identified reason with fire. The soul for him was a fiery particle of the same substance as that of which the heavenly bodies are made, or as the fire on the hearth.

Heraclitus was greatly impressed by his own profundity. He liked to deliver himself of his thoughts in an oracular style, and he deposited the book he wrote in the temple of Artemis at Ephesus so that the intelligentsia who were prepared to wrestle with its obscurities might come there and be made wise. But posterity has warmly endorsed his judgment on himself. The oracle exists now only in some few score of brief disjointed utterances, but it is still worth consulting.

CHAPTER III

EARLY GREEK SCIENCE

PART II : THE ITALIAN SCHOOLS

OUR last chapter was concerned with the history of Greek scientific thought in its first home in Asia Minor. In the generation between Anaximander and Heraclitus refugees fleeing before the advance of Persian power carried Greek thought to the west where it took root in various centres of Greek life in Italy and Sicily. The towns of Croton, Elea, and Acragas successively become the theatres of its most active development.

In its new home in Italy philosophy developed on very different lines from those it had followed in Miletus. It became at once more mathematical and more religious. Pythagoras, the founder of the first Italian school at Croton, is a somewhat mythical figure, but it is clear that the movement he initiated was definitely religious in character, and that the members of his school made astonishing progress in arithmetic and geometry. It was the persuasion of those who came under Pythagorean influence that God, as Plato afterwards put it, is always the geometer.

Pythagoras, according to the most reliable account,

was an exile from the island of Samos. The religious ideas he brought with him to Italy were characteristic of his generation in the islands and coastal towns of the Greek fringe of Asia Minor, and it is customary to associate them with the revival of spirituality brought about by the menace of the Persian advance. Heraclitus, too, shows their influence. He taught a doctrine of the immortality of the soul. For him, as we have seen, the soul, or reason, in man was a portion of the ever-living fire ; and in a passage of his writings that has survived he expounds the religious implication of his system in the following words : " Life and death are present both in our life and death ; for when we live our souls are dead and buried in us, but when we die our souls revive and live."

In the Pythagorean brotherhood also the doctrine that the body is the tomb of the soul was current ; transmigration of souls was taught, and various ascetic practices were inculcated which were designed to keep the soul pure from contamination by its entombment in the body. With this aspect of Pythagoreanism we are not here directly concerned, but it is important to bear in mind that with the Pythagoreans mathematics always bore something of the nature of a religious exercise. The properties of number and space excited in them a mystical awe, and to study mathematics was to imitate the activity of the geometer god.

Greek mathematics, like Greek natural philosophy, begins with Thales. He is said to have introduced Egyptian geometry into Greece and to have developed it further himself. He is credited with having devised a method of determining the height of a pyramid by measuring its shadow, and a method of calculating the distance from the coast of ships at sea. He is also credited with having *proved* that a circle is bisected by its diameter. This is a point of great interest, for it is claimed as the distinguishing quality of Greek geometry that the Greeks, in contradistinction to the Egyptians and Babylonians, first laid down the conditions that require to be fulfilled in order to establish a general proof of a mathematical truth. Thales is also credited with having known, but not with having proved, that the angles at the base of an isosceles triangle are equal, and that if two straight lines intersect the vertically opposite angles are equal.

The cosmology of Anaximander also was not wholly innocent of mathematics. He endeavoured to utilize number and measurement for the understanding of nature. Thus he gave the distances of the wheels of the sun and moon from the earth as multiples of the diameter of the earth; and the construction of his map of course involved some effort to determine distances on the surface of the earth.

With the Pythagoreans, however, mathematics

underwent an extraordinarily rapid development. By the middle of the fifth century they had arrived at most of the results which are set out in an orderly sequence in Books I and II and Books VII to IX of the *Elements* of Euclid. The most famous theorem of Book I, that the square on the hypotenuse of a right-angled triangle is equal to the sum of the squares on the other two sides, is traditionally ascribed to Pythagoras himself; and it is the usual view that a general proof of this important proposition was not forthcoming before the Greeks, although there are those who claim its solution centuries earlier for the Indians and the Chinese. Book II of Euclid sets forth the Pythagorean teaching on equivalent areas, and expounds the method of transforming an area of one form into an area of another form with the consequent possibility of applying areas to one another. This is equivalent to a geometrical algebra of an advanced sort.

The arithmetical discoveries of the Pythagoreans, which provide much of the substance of books VII to IX of Euclid's *Elements*, are of no less interest. After definitions of unit and number there follows a classification of numbers into Odd and Even, with sub-varieties of both, next into Prime and Secondary, then into Perfect and Friendly, after which comes the discussion of Figurate Numbers, Triangular, Square, Polygonal and Oblong. Of Pythagorean origin also is the theory of proportion, the discussion of arithmetic, geometric, and harmonic means, and to some extent the theory of irrational numbers, the knowledge of which by the middle of the fifth century had probably not advanced beyond the recognition of the irrationality of $\sqrt{2}$.

The bulk of numerical theory ascribed to the Pythagoreans is so extensive that it is hardly possible here to do more than indicate its contents in this summary way; but an illustration may be added of the use of Figurate Numbers. They were a device for dealing with the summation of series in a half-geometrical way. The Triangular Numbers, for example, were set out as follows :

$\bullet$, $\overset{\bullet}{\bullet\bullet}$, $\overset{\bullet\bullet}{\bullet\bullet\bullet}$, $\overset{\overset{\bullet}{\bullet\bullet}}{\bullet\bullet\bullet\bullet}$ = 1, 3(1 + 2), 6(1 + 2 + 3), 10(1 + 2 + 3 + 4), etc. That is to say the series of triangular numbers give us the sum of the numbers of the natural series of integers starting from unity. In modern form,

$$1 + 2 + 3 + 4 \ldots + n = \frac{n(n + 1)}{2}.$$ Similarly,

the series of square numbers are equivalent to the formula $1 + 3 + 5 + 7 \ldots + (2n - 1) = n^2$.

In the pages of Euclid the mathematical doctrine of the Pythagoreans is presented to us in severely scientific form. But in the fragmentary remains of Pythagorean writings of early date and in Pythagorean commentaries of later antiquity their mathematics, solid as its content is, is seen to rest upon a curious view of number which warns us that their doctrine existed for its originators in a mental context which it requires an effort for modern minds to reconstruct. To the early Pythagoreans the abstract quality of number was not clear. They are credited with having divorced mathematics from practical ends, that is, with having transformed it from a practical art, as it was in Egypt and Babylonia, into a liberal art. And this is true. But mathematics was far from being for them what it became later, in the famous words of the poet Wordsworth—

> . . . an independent world,
> Created out of pure intelligence.

On the contrary their mathematics was indistinguishable from their physics. We are told by Aristotle, in his own technical terms, that they regarded number not only as the *form* but as the *matter* of things. For them things were numbers. Their arithmetic passed insensibly into geometry, as we have seen in the case of Figurate Numbers, and their geometry passed insensibly into physics. For them a point had bulk, a line had breadth, and a surface thickness. Out of points with bulk they built up lines, out of lines with breadth they built up surfaces, and out of surfaces with thickness they built up solids.

The Pythagorean cosmology bore a general resemblance to that of Anaximander, though naturally the importance of number received full, indeed excessive, recognition. " The whole heaven," they said, " is harmony and number." They were the first to teach the sphericity of the earth and the sphericity of the heavenly bodies, and they dislodged the earth from the central position in the universe. According to them there was a central fire, the hearth of the universe, and round it revolved the earth, the moon, the sun, the five planets, and the heaven of the fixed stars. They imagined also a tenth body, the antichthôn or counter-earth, forever invisible to us because the face of the earth we inhabit is turned away from it. Aristotle says that they invented the counter-earth just to bring the number of the heavenly bodies up to ten, which was a sacred number with them ; modern writers suggest that they hoped thereby to account for the phenomenon of eclipses. One of the major discoveries of the ancients showing an early interest in physical experiment was the Pythagorean discovery that the pitch of notes produced by a taut string is strictly related to the length of the vibrating medium. This discovery was boldly applied to the structure of the universe. It was imagined that the various bodies that revolve round the central fire were fixed at intervals from the centre corresponding to the intervals in the musical scale, and that as they wheeled round eternally they produced a glorious diapason, the music of the spheres, to which our ears, owing to familiarity, are unhappily deaf.

In this fanciful cosmology are embodied ideas that proved to be of the utmost importance. The abandonment of the geocentric hypothesis is in accordance with the Copernican astronomy, and the doctrine of the sphericity of the earth is an advance on the Ionians. Even Anaximander had held the view that the earth was the shape of a flat cylinder. For this advance in cosmology enormous praise has been accorded the Pythagoreans by some historians. The true history of science, however, should be rather a history of method than of results, for the latter are often accidental and only seem impressive to later generations when they have been rediscovered by improved methods. The method of the Pythagoreans was very defective. With the distinction between mathematics and physics non-existent in their minds, and convinced that mathematics is the key to the divine nature, they argued out of a tangle of ethical, mathematical, and physical ideas somewhat as follows : the circle and the sphere are the " perfect " figures, therefore the earth and the celestial bodies must be spheres and move in circles. The defect of this argument is that the earth is not a perfect sphere and that the heavenly bodies do not move in circles, and the true shapes and the true motions are only to be discovered by observation and not by *a priori* mathematical arguments. Subsequently, the Pythagorean method led to the most disastrous results. When it began to appear that nature is indifferent to Pythagorean mathematics and quite careless of shaping her bodies or moving them according to standards of mathematical perfection, those who followed the Pythagorean tradition cast nature off and clung to mathematics. Mathematics became the master instead of the servant. Prostrate before the god who is always a geometer men learned to despise nature which is so careless in the use of the compass and the ruler. It is a form of idolatry which dominated European thought for centuries and for which man has paid a heavy price.

The mathematical view of the universe received its first great shock by the discovery of the irrational nature of $\sqrt{2}$. It was known by the middle of the fifth century B.C. that the side and the diagonal of the square are incommensurable, and the discovery was the occasion of a crisis in Pythagoreanism. The thing became a scandal. As the very word " irrational " implies, reason seemed to be threatened by it. The whole heaven was harmony and number, but here was a simple relation that could not be expressed by any number. Tradition even reports that the Pythagorean brotherhood tried for a while to suppress the scandal by concealing the new discovery, a melancholy precedent often repeated.

The discovery of the irrationality of $\sqrt{2}$ brought the edifice of Pythagorean number-physics down with a crash. If the side and the diagonal of the square are incommensurable it means that no matter how long you go on dividing up these lines you will never get to integers in which you can express the relation of one length to the other. The conclusion follows that lines are divisible to infinity. But if lines are infinitely divisible they do not consist of a countable number of points. Now these Pythagorean points were the little bricks of which the universe was built, and they had dwindled to nothing. The foundations of the Pythagorean universe had been swept away.

The relentless exposure of the collapse of the number-physics of the Pythagoreans was the work of Zeno, a member of another Italian school, that of Parmenides of Elea, to which we must now turn. Before we examine the criticism of Zeno we must first understand the strange and brilliant theory of his master Parmenides.

Parmenides, the founder of the Eleatic school, was a contemporary of Heraclitus of Ephesus, to whose teaching of the Upward and the Downward Paths he makes allusion. This is interesting proof of the traffic in ideas between the eastern and the western Greeks at the end of the sixth century. The doctrine of the Eleatic school is the creation of a man in touch with both Ionian cosmology and Pythagorean number-physics. Parmenides, like Heraclitus, was impressed by the unreliability of the evidence of the senses and exalted the reason above the senses as the organ by which we apprehend truth. But he came to a different conclusion on the nature of things from that which commended itself to Heraclitus. Heraclitus maintained that everything was in a state of flux ; Parmenides denied absolutely the possibility of motion and change. Motion and change, he averred, were illusions of the senses.

Nobody at this early date had yet imagined that there might be such a thing as an immaterial existence. It will be remembered that even Heraclitus had identified reason with fire. Parmenides accordingly denied the existence of empty space. He said : That which is, is ; that which is not, is not. Therefore matter, or the stuff of which the universe is made, cannot be thinned out or differentiated in any way. It must be absolute fullness of Being without any quantitative or qualitative diminution of its fullness. The universe, he taught, is one solid uncreated indestructible motionless changeless sphere. Like God, with Whom indeed Parmenides identified it, it is the same yesterday, to-day, and for ever.

For the idea that what exists is a sphere Parmenides is indebted to the Pythagorean notion of the perfect figure. But in his rejection of motion and change he criticizes both Pythagorean and Ionian physics from the standpoint of a stricter logic.

All the Ionian philosophers had taught that matter passes from one form into others, and the mode of this transformation had been best explained by the theory of Anaximenes of Rarefaction and Condensation. According to this view what differentiates one form of matter from another is simply the amount of it that is packed into a given space. But if there is no such thing as empty space, there cannot be more or less of matter in a given space. Space must be absolutely full, and the manifold variety of this changing world must be an illusion of the senses. The truth for reason is the changeless One, and if we seek wisdom it is to the One we must cling.

This strange philosophy, affirming the One which is true for reason and rejecting the Many, which our senses perceive, was the result of a stricter logic than had yet been applied to the problem of Being. Parmenides, who was the first philosopher to express himself in verse, sang his message with impressive power, appealing to the " deaf, blind, bewildered multitudes who could not distinguish Being from Not-being." Shelley, in a famous stanza in *Adonais*, repeats the strain :

> The One remains, the Many change and pass ;
> Heaven's light for ever shines, earth's shadows flee ;
> Life, like a dome of many-coloured glass,
> Stains the white radiance of eternity,
> Until Death tramples it to fragments.

The Pythagoreans had warred against the senses because they looked upon the body as a perishable thing in which an immortal spirit was entombed ; but in their number-physics that had attempted to give an intelligible account of the manifold world of the senses. Here was a new philosophy which in the name of reason denied the reality of the whole world of the senses, maintaining that the only reality was the static unity of a spherical universe identical with God.

This new doctrine found itself at variance with common sense, Ionian cosmology, and Pythagorean number-physics. To support the Parmenidean view a young disciple, Zeno, about the middle of the fifth century, devised an ingenious system of paradoxes designed to show that any departure from the Parmenidean One landed the heretic in hopeless contradictions. The most famous of these paradoxes is that known as the paradox of Achilles and the tortoise. It is the only one we have space to examine.

Zeno presented his problem as follows : Achilles and a tortoise are to run a race ; Achilles runs ten times as fast as the tortoise ; the tortoise gets ten yards start. Experience tells us that Achilles will quickly pass the tortoise, and common sense finds it quite simple that it should be so ; but a little logic will reveal that the common sense view is not so simple as it appears. Let us try to examine the stages of the race. Achilles must first run ten yards to the point from which the tortoise started. Meantime the tortoise runs one yard. Achilles runs that yard. The tortoise meanwhile advances one tenth of a yard. While Achilles does that tenth the tortoise does a hundredth, and so on. The distance between the two constantly diminishes, but Achilles never gets past the tortoise. The common sense view of motion, then, which experience seems to endorse, is nevertheless irreconcilable with logic. It is better to stick to logic and reject the muddled, sense-begotten, notion of motion.

The argument against the reality of motion was put still more succinctly in the following way : " There is no motion, for that which is moved must arrive at the middle of its course before it arrives at the end." This argument rests upon the idea of the infinite divisibility of space. Between any two points lies an infinite number of points. If a thing begins to move at all, it has already moved an infinite distance. The very idea of motion is self-contradictory.

The purpose of these arguments will be better understood if we take them in connection with the crisis in Pythagorean number-physics which we have already discussed. It will be remembered that the Pythagoreans had attempted to build up a universe out of points with magnitude. The discovery of the incommensurability of 1 with $\sqrt{2}$ had forced upon them the recognition of the infinite divisibility of space. The paradoxes of Zeno were devised to meet this situation and to reveal the completeness of the dilemma which had overtaken the number-physics. On the Pythagorean view that the universe is made up of a plurality of units it can now be made to appear that the universe, or any portion of it, is both infinitely great and infinitely small. For suppose that when we have carried the process of division to infinity we arrive at a point of some magnitude, then the total must consist of an infinite number of units all of some magnitude, and must therefore be infinitely great. If on the other hand the process of division brings us at last to units of no magnitude, then an infinity of units of no magnitude still amounts to nothing at all. The Pythagorean number-physics had no escape from this dilemma.

The arguments of Zeno have sometimes been belittled as ridiculous fallacies and sometimes praised as if their profundities were still but half plumbed. A less extreme view is to recognize that against the Pythagorean number-physics they are valid, but that once we abandon the Pythagorean view of space as made up of a series of discrete points their applicability ceases. But the name of Zeno remains a great one in the history both of logic and mathematics. In his large *History of Greek Mathematics*, Sir Thomas Heath remarks:

E

" It would appear that, after more than 2,300 years, controversy on Zeno's arguments is yet by no means at an end." All we can hope to do here is to indicate some of their consequences for the subsequent development of Greek thought.

In the sphere of mathematics their effect was to banish number from geometry. The arithmetical geometry of the Pythagoreans had reached its limit. A great mathematician, Eudoxus, devised a technique of representing ratios of magnitude geometrically, which was applicable to all magnitudes, whether commensurable or not. This type of geometry was taken up by Plato and developed by Euclid. It is a geometry in which lines are no longer regarded as being built up of points, and surfaces of lines, and solids of surfaces, but in which spatial relations can be symbolized quite independently of number and studied without reference to measurement. This is the independent world created out of pure intelligence, the mathematics a training in which Plato looked upon as the necessary preliminary to philosophy.

Arithmetic, too, underwent a change. It was subdivided into an abstract science known to us as Theory of Numbers, the arithmetic proper of the Greeks, and *logistic*, or the art of calculation, which was separated from the theory of numbers and fell into neglect among the exponents of a liberal education as being contaminated by practical uses.

It was felt to be a proper study for Phœnician traders. The term " calculator " was still a term of abuse with Edmund Burke in the eighteenth century.

But the greatest effect wrought by the arguments of Zeno on Greek thought was in the sphere of physical speculation, and this will form the subject of our next chapter. Obviously, after Zeno's criticism, physical speculation had to relapse into contemplation of the Parmenidean One or justify a world of motion and change by profounder arguments than had yet been used. The first effort to escape the paralysis of the Parmenidean One and restart speculation on a pluralistic basis admitting motion and change came from the Sicilian school of Acragas and was the work of its most famous citizen Empedocles.

CHAPTER IV

THE ATOMIC THEORY

By the middle of the fifth century Greek physical speculation had begun to exhibit considerable variety both in method and in conclusions. The method of the early Ionians based on the assumption that all the manifold phenomena of nature could be explained in terms of the objects and processes familiar to ordinary observation had been made to seem insufficient in the light of the logical criticism of Parmenides and the mathematical physics of the Pythagoreans ; while their guesses that the primary substance could be identified with water, or the indeterminate, or mist had likewise been judged inadequate. Heraclitus, though he had chosen a new primary substance, fire, had put the main emphasis on his theories of the flux and of opposite tension. The Pythagoreans had stressed the importance of the pattern of things, of numerical and geometrical relations. And Parmenides, in the name of reason, had dismissed the world of sense as an illusion.

On the general life of Greece at this time Pythagoreanism, with its cult of number and geometry,

made the widest impression. The sculptor Poly-
cleitus, for instance, like many artists in later days,
fell a victim to the prevailing intellectualism. He
no longer saw things with the innocent eye of the
artist ; but he began to detect mathematical proportions in the human body, wrote a book to
expound the new philosophy of art, and even made
a statue, the famous Doryphorus or Man with a
spear, to illustrate his thesis. The most interesting
practical application of Pythagoreanism, however,
lay in another field.

If the whole heaven was number and harmony,
it was fitting that the earth should be made as far
as possible to conform. Town-planning on geometrical lines became the fashion. Hippodamus
of Miletus, an enthusiast for the number philosophy, made new plans for Peiræus, the port of
Athens, for Rhodes, and for Thurii. In the next
century the towns of Alexandria and Priene show
the persistence of his influence. The practice
descended to Roman times. The lay-out of
Pompeii and Timgad, to choose two examples out
of many, are later testimonies to the Pythagorean
tradition ; and many modern towns which have
sprung up out of Roman camps still exhibit the
impress of this fashion. Even the New World is
heir to the tradition. New York, with its geometrical lay-out and its numbered streets and avenues,
is a thoroughly Pythagorean town.

But while the practical results of the advance of knowledge were thus leaving their mark on the material civilization of the ancient world, science itself in its endeavour to find out the true path of progress was facing a great crisis. The doctrine of Parmenides had brought the debate between reason and the senses to a head. Along which route did the path to knowledge lie ? Parmenides himself had no doubt. Reason was the true guide and the evidence of the senses was to be rejected. In the prelude to his philosophical poem the attack on the senses is formally opened. " Turn your mind away from this path of enquiry," he writes. " Let not the habit engrained by manifold experience force you along this path, to make an instrument of the blind eye and echoing ear and the tongue (the organ of taste), but with your reason weigh my contribution to the great debate." It was not possible to ignore the challenge ; the problem of determining the validity of sense-perception had to be attacked.

It was attacked in the most direct manner by Alcmæon of Croton who endeavoured by anatomical research to expose the physical basis of sense-perception. This brilliant early anatomist, the founder of empirical psychology, was, as was natural in an inhabitant of Croton, an adherent of the Pythagorean school. Although the Pythagoreans are chiefly famous for their mathematical theories it must not be supposed that they were indifferent to the claims of physical investigation. It is often stated that the ancients were innocent of all understanding of the method of experiment. This is not true. In their successful demonstration that the pitch of the note produced by a taut string depends on the length of the vibrating medium the Pythagoreans had utilized a movable bridge to vary the length of the string, thus applying precise measurement to the investigation of a physical phenomenon. And the Ionians have also left proof of their addiction to the practice of observation and experiment. The cosmological speculations of Anaximander plainly had their basis in observation, and the astronomical activities of his school were subsequently carried on at a suitable observation post in the island of Tenedos. Anaximenes also appealed to experiment in support of his theory of rarefaction and condensation. He pointed out that if we open our mouths wide and emit rarefied air against our hand it feels warm and is therefore presumably on its way to being transformed into fire, while if we condense the air by forcing it through pursed lips it feels cold and is therefore presumably on its way to being transformed into the cold element of water. His conclusions are wrong. Anaximenes did not know that the sensation of coldness is due to the evaporation of moisture on the surface of the skin. But the method is that of experiment.

There is nothing, then, to surprise us in the fact that when the nature of sense-perception became a problem Alcmæon should have sought to settle it by resort to dissection and vivisection. In the course of his investigation he discovered the optic nerves and came also to the correct conclusion that the brain is the central organ of sensation.

These researches of Alcmæon exhibit an inclination neither to accept or reject uncritically the evidence of the senses but to determine experimentally the limits of its validity. This was also the point of view of Empedocles, the philosopher of Acragas, who endeavoured, while allowing for the Parmenidean criticism, to reconstitute the Old Ionian tradition on a sounder basis. Empedocles had quite an elaborate theory of the nature of sense-perception, to which we shall return later ; for the moment all that concerns us is to realize that he decidedly rejected the appeal of Parmenides to dismiss sense-evidence as wholly misleading. Empedocles, like Parmenides, wrote in verse, and in a passage of his poem which has survived he replies to Parmenides in the following words : " Go to now, consider with all thy powers in what way each thing is clear. Hold nothing that thou seest in greater credence than what thou hearest, nor value thy resounding ear above the clear instruction of thy tongue ; and do not withhold thy confidence in any of thy other bodily parts by which there is an opening for understanding, but consider everything in the way in which it is clear." The attitude of Empedocles is very much that which has presided over the growth of science down the centuries. He laments the shortness of human life and the fallibility of the senses but thinks that with patience and caution a true knowledge of nature will accumulate from generation to generation.

Empedocles had thus cleared away one of the obstacles raised by Parmenides across the path of natural philosophy. But there remained the formidable difficulty of the Parmenidean One. The logic of *What is is ; what is not is not* seemed to make it impossible to restore variety, multiplicity, movement, and change to the world so long as philosophers insisted on seeing at the basis of all existence one primary substance. Empedocles decided to abandon the tradition of Monism, as it is called, and to reconstruct his universe on the assumption that there are not one but many primary principles. In this way, by putting multiplicity and variety back into the essential nature of Being, he sought to justify logically the manifold, changing world of sense-experience.

According to Empedocles there are four primary substances, or " roots " of things, earth, air, fire, and water, what we still to-day in popular speech call the four elements. These elements, of course, were borrowed from his Ionian predecessors. He had also to provide some machinery for putting the elements in motion, some equivalent for the rarefaction and condensation of Anaximenes and the Upward and Downward Paths of Heraclitus. This he supplied by two forces which he called Love and Hate, Love which tended to unite the four elements into a mixture and Hate which tended to draw them apart again. Since nobody had as yet distinguished force from substance Empedocles regarded love and hatred as being material things which formed part of the mixture. To historians of a generation or two back this inability to conceive of a force except as a corporeal substance seemed proof of the childishness of Greek thought at this epoch. Now that the distinction between matter and energy has ceased to be fundamental it may possibly occur to some historian to credit Empedocles with a profound insight into the most recent development of physics. Neither view is correct. The important thing is to remember that no scientific concept can be properly understood out of its historical context, and all the early Greek thinkers were materialists.

Accordingly, the psychology of sense-perception which, as we have already remarked, Empedocles treated with some elaboration, was put by him on a completely materialistic basis. His idea was that we are composed of the same elements, Earth, Air, Fire, and Water, and the same forces, Love and

Hate, as the rest of existing things, and that sense-perception is of the nature of physical intermingling of like elements. By fire we recognize fire, by water water, and so on. But whether two drops of water " recognize " one another when they meet he did not say, nor if not, why not. These questions belong to a later generation.

The biological speculations of Anaximander were also taken up and developed by Empedocles. References to his theories on the origin of different forms of life survive in fragments of his poem and in the commentaries of later writers. His ideas are reproduced also, though with some modifications, in the *De Rerum Natura* of the Latin poet Lucretius, from whom the following passage, in the version of Munro, is quoted : " And many monsters too the earth (in early days) essayed to produce, things coming up with strange face and limbs, the man-woman, a thing between the two and neither the one sex or the other, widely differing from both ; some things deprived of feet, others again destitute of hands, others too proving dumb without mouth, or blind without eyes, and things bound fast by the adhesion of their limbs all over the body, so that they could not do anything nor go anywhere nor avoid the evil nor take what their needs required. Every other monster and portent of this kind she would produce, but all in vain, since nature set a ban on their increase and they could not reach the coveted flower of age nor find food nor be united in marriage. For we see that many conditions must meet together in things in order that they may beget and continue their kinds ; first a supply of food, then a way by which the birth-producing seeds throughout the frame may stream from the relaxed limbs ; also in order that the woman may be united with the male, the possession of organs whereby they may each interchange mutual joys.

" And many races of living things must then have died out and been unable to beget and continue their breed. For in the case of all things which you see breathing the breath of life, either craft or courage or else speed has from the beginning of its existence protected and preserved each particular race." (Bk. V, 837–859.)

These arguments, amounting as they do to a rough forecast of the Darwinian theory of Natural Selection, deserve mention in even the briefest account of Greek speculation.

Before we leave Empedocles mention must be made of one brilliant discovery, dependent on experiment, which stands to his credit. Before him no physicist had clearly distinguished air from empty space. He was the first to demonstrate conclusively that the viewless air is a bodily substance. The instrument that he used for his experiment was the clepsydra or water-clock. The clepsydra was a metal cylinder terminating in a cone. At the tip of the cone was a small hole ; the base of the cylinder at the opposite end from the cone was formed by a strainer with small perforations. The instrument was filled by thrusting the broad end under water ; it was put into use by inverting it and letting the water escape through the small hole at the tip of the cone. The water emptied itself out in a fixed time ; and clocks of this type were in regular use at a later date in Greek lands, especially in law-courts where they served to control the time allowed for the speeches of the rival sides. In the time of Empedocles it is possible that the clepsydra, which literally translated means " water-stealer," was known only as a toy. The two experiments by which he demonstrated the corporeal nature of air were as follows. In the first the broad end of the clepsydra was thrust under water in the usual way for filling it, but a finger was kept tight over the hole at the tip of the cone. No water entered since the enclosed air could not escape. In the second demonstration the clock was filled in the usual way, the opening in the tip was closed, and it was lifted out of the water without being reversed. In this experiment no water could escape owing to the pressure of the atmosphere outside. Later experimenters reinforced this demonstration of the corporeal nature of air by inflating bladders and exhibiting their resistance to compression.

The experiments of Empedocles on air and the Pythagorean experiments with musical strings are the best examples of physical experimentation that have come down to us from the early period of Greek science. Meantime, as we have seen, physiological investigation had been begun on a scientific basis by Alcmæon of Croton, and this movement had its parallel also among the Asiatic Greeks, as we shall see later. The interest in physiological studies strongly colours the views of the next thinker of importance in the history of Greek science. This was Anaxagoras of Clazomenæ who settled in Athens in the middle of the fifth century and as a member of the brilliant circle that clustered round the great Pericles first brought the mainland of Greece into touch with the movement of speculative philosophy which had developed among the Greeks of Asia and Italy.

Like Empedocles, Anaxagoras abandoned monism ; he is indeed distinguished as the propounder of the most thorough-going pluralism that could be conceived. According to him the first principles or " seeds " out of which the universe is made are infinite in number and in variety, and every one of these seeds contains a part of every substance of which our senses give us knowledge. This strange theory had its roots in physiology. The growth of vegetable life by elements drawn from the earth, the growth of animal bodies by the consumption of vegetable life, were problems that deeply interested Anaxagoras and gave a new and fruitful turn to the old speculation on the process by which things apparently change into one another. If a man eats bread the result is that he grows flesh, blood, bones, sinews, skin, hair and so forth. How, asked Anaxagoras, could this be possible unless bread somehow contained flesh, blood, bones, sinews, skin and hair ? Arguing on these lines he came to the conclusion that the ultimate elements out of which the universe is composed each and all contain a portion of everything. Things were infinitely divisible according to him, but however far you carried the division you would never come to a portion of matter so small that it did not contain a portion of everything. It is obvious that Anaxagoras is groping after the idea of a union that would be much closer than that of a mere mixture or physical juxtaposition ; and while it would be too much to say that he had arrived at the idea of chemical combination, he had certainly gone some of the way towards it. To complete the outline of his theory of the composition of matter it remains to add that though he thought that every seed contained a portion of everything, he did not suppose that all the seeds contained equal proportions of everything. In the various groups of seeds some one thing predominated. Thus water would be made up of seeds that consisted predominantly of water, while containing portions of everything else, and so for all the other substances.

Empedocles' doctrine of the four elements or roots of things, and Anaxagoras' theory of the primary seeds each of which contains a portion of everything, are proof of the vigorous intellectual effort devoted at this time to the solution of the problem of the constitution of matter. Neither of these solutions, however, has excited or merited so much admiration as the atomic theory of the constitution of matter propounded by Leucippus and Democritus in the second half of the fifth century. This marks the close of the great movement of physical speculation instituted by Thales.

The credit for the first suggestion of the atomic theory must go to Leucippus. He is said to have been a native of Miletus and to have spent some time in contact with the Eleatic school ; the tradition has at least a symbolic truth, for the function of Leucippus was to develop the viewpoint of the Milesian school while allowing full weight to the criticism of Parmenides. Leucippus felt that Parmenides was absolutely right in insisting that the primary substance was solid, uncreated, indestructible, motionless, changeless, uniform in its essential nature, and possessed of absolute fulness of being. In this sense it was One, as Parmenides taught. But he refused to accept the doctrine that it was continuous. On the contrary, it existed in the

form of minute particles far too small to be perceptible by our senses. These particles of matter, the
atoms, infinite in number, were separated from one
another by void. It was in introducing the doctrine
of void that Leucippus broke with Parmenides. Of
the two famous propositions of Parmenides, *What
is is, what is not is not,* Leucippus accepted the first
and denied the second. According to him matter
exists and void also exists. It is the nature of
matter, as Parmenides taught, to be completely full,
completely impenetrable, in our phrase. But there
is also another mode of existence, that of void,
which consists in being completely empty, completely penetrable. Out of these two existences,
atom and void, it is possible, Leucippus thought,
to construct a universe which will satisfy both
the claims of logic and common sense.

Logic demanded that there should be some one
permanent substance underlying the world of
change. Common sense demanded that the plain
testimony of our senses as to the existence of a
manifold and changing world should not be sacri-
ficed to the demands of logic. The doctrine of
Leucippus satisfied both demands. The atoms, all
uniform in substance, were exempt from change;
but the combinations of atoms which made up the
visible, tangible world, were always coming into
being and perishing. The coming into being of
any object of sense was the result of a grouping of

atoms, the change of any object of sense was the regrouping of atoms, the passing out of existence of any object of sense was the dispersal of the atoms ; but the atoms in themselves neither came into existence, changed, nor perished. The atoms, all alike in substance, differed from one another in size, shape, and arrangement. All other differences in perceptible things were merely the effects on our senses of atoms of different shapes and sizes grouped in different ways. Thus the world of the senses was not absolutely primary in the way that atoms and void were, but neither was it an illusion.

This brilliant theory superseded all that had preceded it. Parmenides was provided with the unchanging substance for which he had asked. The earth, air, fire, and water of Empedocles and the old Ionians were allowed reality, but were all explained as products of the primary atoms grouped in different ways. The difficulty of Anaxagoras as to how bread could become blood, flesh, and bone was provided with a ready answer. And to the ingenuity of the Pythagoreans in suggesting patterns on which the visible universe is constructed a boundless field was opened up, which, however, remained unexplored until the birth of modern physics and chemistry. To the problem of the constitution of matter Leucippus provided the best answer antiquity was able to give.

It is probable that it is to Democritus, the disciple

and associate of Leucippus, that we owe the working out of a new cosmology on the basis of the atomic theory of matter. This man was a native of Abdera in Thrace, and has had the misfortune to become known to posterity under the absurd title of the laughing philosopher, owing, it is said, to the imperturbable cheerfulness which characterized him throughout a long life. He was a voluminous writer on every branch of science then cultivated, and the loss of his works is probably the most serious in the all but universal ruin that has overwhelmed the record of the earlier thinkers among the Greeks. He is notable for his comprehensive grasp and superb powers of generalization. As the basis of his system Democritus laid down the principle : " Nothing is created out of nothing or destroyed into nothing." The proposition was not new, but Democritus was the first to put it in its proper place as the first principle of all scientific thought about the physical world. He further showed his wisdom by treating it as an axiom and offering no proof. The only possible proof of such a proposition is the pragmatic one. It is true in so far as it works.

Having first posited the doctrine of the permanence of matter Democritus went on to enunciate the law of the universality of cause and effect. " By necessity were foreordained all things that were and are and are to be." This is the first clear enunciation in the history of thought of the principle of determinism. In the light of this principle science is seen as a knowledge of causes, and the goal of the physicist is the discovery of the necessary sequence of events.

Having postulated the doctrines of the permanence of matter and the universality of the reign of law, Democritus proceeded to explain the origin and operation of our world, or of any world (for the atomists did not believe in the uniqueness of our world, but thought that worlds are infinite in number and are always coming into existence and passing away), on the basis of the new theory of the constitution of matter. The elements of which a world are made are two, atoms and void. The atoms are conceived of as being naturally in a state of violent agitation in the void. As they collide with one another and rebound from one another and impart their motions to one another, it will happen that the larger atoms will tend to cluster together, since an impact with another atom does not drive them off to so great a distance as in the case of a smaller atom. The massing together of the larger atoms produces the heavier earthy substances which lie at the centre of a world. Outside this mass, which is supposed to be in circular motion as the result of the collisions of the atoms, cluster smaller atoms and of a shape less adapted to close cohesion ; these form first water, then air, and then fire, which consists of the minutest atoms of all. But it is natural to suppose that in the outer sphere of finer atoms some larger atoms should remain which at first form earthy and moist lumps, much like the earth. But owing to the rapidity of their motion in the whirling sphere they first dry and then ignite, and so form the sun, moon, and stars.

Obviously generations of research in sciences as yet unborn were necessary before the manifold problems raised by this daring speculation could be tested, accepted, discarded, corrected, supplemented; but it remains one of the supreme achievements of scientific thought. Before we leave it two points may be selected for special emphasis. In the first place the conception of the atom, an indivisible particle of matter, had made a sharp and needful distinction between the domains of mathematics and physics. In the number-physics of the Pythagoreans matter had not been distinguished from spatial extension, and the discovery of the infinite divisibility of space had swept away the basis of their physics. Now the essentially physical concept of impenetrability is made the corner-stone of the edifice of physics; the atom is spatially divisible but physically indivisible. Numbers can no longer be confused with things. In the second place science had raised a certain nightmare for the human spirit. In this vast universe of atom and void, where all was under the reign of mechanical law and where worlds formed and burst like bubbles in a stream leaving not even a memory behind, what room was there for human will and human effort, for human hopes and human fears? The imperturbable Democritus does not seem to have felt the difficulty, for he is credited with the composition of ethical as well as physical treatises. But subsequent generations of thinkers have not ceased to wrestle with the task of reconciling Determinism and Free Will.

CHAPTER V

GREEK MEDICINE

THE SPREAD OF THE SCIENTIFIC SPIRIT

IN all the period that has been discussed so far it has not been profitable, or even possible, to draw a clear line between science and philosophy. Some of the thinkers, like Parmenides and Heraclitus, who rely mainly on their powerful logic, correspond rather to our idea of a philosopher; others, like Anaximander or Democritus, who are steeped in the observation of phenomena, come closer to our conception of a scientist. But all alike, being engaged in speculation on the nature of things in general with but little accumulation of positive knowledge to guide them, and advancing opinions the truth of which could not readily be put to the test of experiment, belong to a stage of development in which the spheres of the scientist and the philosopher have not as yet been separated. With the school of research, however, which forms the first subject of this chapter, we arrive at a clear distinction between the method of the scientist and of the philosopher. It is the most important gain for the method of science that emerges from the work of this school.

This school, the medical school of Cos, is famous in history as being the first scientific institution from which complete treatises have come down to us. Of the productions of the earlier schools, whether of Ionia or Magna Græcia, nothing has survived except fragments embodied in the writings of later generations. But from the school of Cos we have a collection of some thirty odd treatises, put together at a later date into a corpus known as the Hippocratic Collection, but certainly consisting mainly of the work of members of the school. The school was founded about 600 B.C.; the earliest extant work was composed about 500 B.C.

The great man, Hippocrates, after whom the school came to be called, was born about 470 B.C. and lived to a great age, dying perhaps as late as 370 B.C. Of the three or four treatises we shall mention here some were possibly the work of Hippocrates himself, all were contemporary with him and embody his teaching. They are of supreme importance in the history of scientific thought.

Greek medicine derived from a number of sources all of which contributed something to the tradition that established itself at Cos under the presiding genius of Hippocrates. In the first place must be mentioned the old sacerdotal medicine exercised by the priests of Æsculapius, the God of Healing. This would differ from the lay science of the Greeks as being the monopoly of a closed corporation. It probably also contained an element of humbug ; but the success of humbug both in comforting and curing patients is not to be denied. From this sacerdotal source a rich accumulation of experience must have reached the early doctors of the school of Cos, who were themselves priests of Æsculapius but appear to have broken down the barriers of secrecy and exclusiveness that hemmed in the older temple medicine.

A second source on which Greek medicine drew was the physiological speculations of the philosophers. Alcmæon of Croton, whose discovery of the optic nerves and whose theory that the brain is the seat of sensation have been already noted, combined the practice of anatomy with Pythagorean theorizings as to the nature of health and disease. He was of opinion that health is a harmony of opposites. But such vague generalizations are of little use in medicine. If a man is ill it is obvious that the balance of his constitution is upset. But a doctor whose only advice was to restore the balance would not be found very helpful. Another Pythagorean, Philolaus, to whom, as a Pythagorean, the number four held a special importance, taught that there are four principal organs in the human body, the reproductive organs, the umbilicus, the heart and the brain, representing four ascending stages of life. Common to all living things is the power of reproduction ; to this plants add the power of growth, animals a third power, sensation, and man a fourth, reason. Accordingly, man has sexual organs, the seat of his reproductive power ; a navel, the seat of his vegetable life ; a heart, the seat of sensation, which he shares with the rest of the animal world ; and a brain, which is the seat of his rational, or specifically human, life. Here it is clear that an *a priori* schematization has produced an appearance of tidy logic in the human constitution at the expense of truth. The emphasis on the navel to the exclusion of such major organs as the liver, kidneys, and lungs is absurd ; and the choice of the heart as the seat of sensation is not only wrong but arbitrary. Empedocles is a third example of a philosopher who endeavoured to contribute to medical theory. He wrote a treatise on healing in which he sought to apply his doctrine of the four elements to the cure of disease. He taught that there were two pairs of opposite qualities associated with the four elements, namely the wet and the dry, and the cold and the hot. Earth was dry and cold, water was wet and cold, air wet and hot, fire dry and hot. We lack the details of his application of this scheme to therapy, but it must have been somewhat as follows. If a man is shivering and has a running nose, he has too much cold and wet in him and must be cured by hot and dry ; if on the other hand he is in a high fever, administer cold !

The third and last source of Greek medicine

was the most modest and the most important : to wit, the directors of gymnasiums. These men learned to deal with fractures and dislocations, and it is no doubt largely as a result of their accumulated experience that the surgical treatises in the Hippocratic collection are on such a high level. The requirements of the patrons of the gymnasiums also stimulated research in other directions. We hear of one, Iccus of Tarentum, who studied diet, and of one, Herodicus of Selymbria, who applied gymnastics to the cure of disease. It would be impossible to stress too strongly the importance of these lines of experiment and research for the development of Hippocratic medicine. Surgery, diet, and exercise comprise almost all the therapeutic aids of the Hippocratic doctors.

With the Hippocratic school of medicine we enter the domain of science in the fullest sense. Of positive knowledge we must not indeed expect very much. The Hippocratic doctors had no thermometer, no magnifying glass, no stethoscope ; neither had they any anatomical knowledge beyond naked-eye surface anatomy. Their physiological knowledge was consequently almost nil also, for they were ignorant of the functions of the internal organs and had no means of analysing the waste products of a living organism. Nevertheless their knowledge, slight as it was, is fully entitled to the name of science. We find in their writings a clear conception of medicine as based on observation of the behaviour of the human body in health and in disease, on experiment, and recording of results. It is recognized that the fund of medical knowledge can only grow slowly with the passage of the generations, and a tradition is established of teaching the accumulating results of experience. We find this growing body of knowledge defended against superstition, which never affects humanity more powerfully than when the reason is shaken by the terror of disease and death. Still more noteworthy is it that we find the science defended, as an observational and experimental one, against the encroachments of the philosophers who come with their ready-made views of the nature of man derived from cosmological speculation and attempt to base the practice of medicine upon them. Thus for the first time a clear distinction is drawn between the nature of an observational and experimental science and the speculation that is permissible in dealing with a material inaccessible to the experimental method. And it is finally to be noted that the Hippocratic doctors held an exalted ethical ideal of their profession as a disinterested service of humanity. These various ideas found expression in aphorisms current in their school. Emphasizing the truth that science is a slow growth demanding the co-operation of successive generations, they said : *Art is long, life is short.* And feeling the sacredness of the office they were called upon to fulfil they said : *Where the love of mankind is, there is the love of the art.*

Among the more notable of the treatises in the Hippocratic writings is a collection of Case Histories. These record the clinical observations of the physician throughout the whole course of various illnesses, some lasting several weeks. The different symptoms and phases of the diseases are recorded with a business-like accuracy ; the treatment, which generally amounted to no more than keeping the patient in bed and feeding him on slops, is set forth ; and the result, which in the majority of cases was death, faithfully reported. These case histories are absolutely free from the slightest touch of superstition ; and in another treatise, called *On the Sacred Disease*, we find a formal and explicit denial of the supernatural view of the nature of disease. The Sacred Disease was epilepsy, then generally regarded as a divine visitation. " It seems to me," says the writer, " that the disease is no more divine than any other. It has a natural cause, just as other diseases have. Men think it divine merely because they do not understand it. But if they called everything divine which they do not understand, why there would be no end of divine things." Thus with a combination of gentle irony and sound argument is superstition banished from the realm of medicine.

The writing in which the method of medicine, as a science of observation and experiment, is defined and defended from the specious competition of the philosophers is that known as the *Tract on Ancient Medicine*. The importance of this treatise is so great that we propose to examine it at some length. It opens with a direct attack upon the enemies of the established tradition of medicine, who are defined as " all those who attempt to speak or write about medicine with an hypothesis or postulate as the basis of their argument." The Empedoclean elements, looked at from their qualitative side, the Hot, the Cold, the Wet, the Dry, are then mentioned as examples of such postulates ; and those who attempt to narrow down the causal principle of diseases and death to one or more of these postulates are censured for interfering with an art which is of the utmost importance to mankind and which depends for its progress on research and discovery. The chapter concludes with a sarcastic recommendation to the philosophers to keep their postulates for dealing with insoluble mysteries, " for example, things in the sky, or below the earth. If a man were to learn and declare the state of these, neither to the speaker himself nor to his audience would it be clear whether his statements were true or not. *For there is no test the application of which would give certainty.*"

" As for medicine," the second chapter continues,

" it has long had all its material to hand, and it has discovered both a principle and a method through which many excellent discoveries have been made over a long period, and by pursuing which full discovery will be made, if the researcher be competent and conduct his enquiries with knowledge of the discoveries already made, using them as a starting-place. Whoever rejects these means and attempts to conduct research in any other way is a dupe or a cheat."

The art of medicine, the argument proceeds, came into existence because sick men do not get well if they continue to eat the same food as men in health. By observation of what modifications must be made in food to suit various constitutions and conditions of health, a body of knowledge has grown up which, though incomplete and limited in range, is still true because it can definitely heal some cases and because it can be taught. This is the method, this is the tradition, of " ancient medicine " which must be maintained against the innovations of the philosophers.

The philosophers when they turn to medicine base themselves on postulates. That is to say they bring with them from their speculations on the nature of things some such theory as that sickness results from too much Hot, or Cold, or Moist, or Dry. Necessarily, then, they try to counteract Cold with Hot, or Moist with Dry. Now suppose a man is sick whose normal diet is wheat, raw meat, and water. A practitioner of ancient medicine will treat him by substituting bread for wheat, boiled meat for raw, and adding wine to the water. What remedy would the new philosophical physician suggest? A diet of Hot, Cold, Moist, or Dry? " I think I have non-plussed him," says the writer, allowing himself a modest triumph. He then goes on to point out that there is no such thing as Hot by itself, or Cold by itself, and so forth ; there are only hot, cold, wet, and dry *things*, and what the doctor must prescribe for his patient is some *thing* in which various qualities inhere ; and of these qualities the hot, the cold, the moist, and the dry are not the most important. For on the one hand the human body has an internal capacity of reacting to temperature ; and on the other hand it is such qualities as bitterness, sweetness, acidity that disturb the health of man when they exist in excess. Diseases are the result not of one or two but of a vast variety of causes ; and " we must surely consider the cause of each complaint to be those things the presence of which necessarily produce a complaint of a specific kind, and in the absence of which it ceases to exist."

Such is the argument of this brilliant treatise. It is probably the soundest exposition of the observational and experimental method that has survived from antiquity and it played its very definite part in the recreation of scientific medicine in the modern era. Thus Jacques Massard, Dean of the College of Physicians at Grenoble and a member of the (French) Royal Academy for New Discoveries in Medicine, writing in 1686, acknowledges its worth and its influence. " Hippocrates himself," he tells us, " composed this treatise on Ancient Medicine to oppose certain innovators of his time who were endeavouring to establish as the cause of diseases, the hot and the cold, the dry and the moist, and by this false principle were disturbing the foundation of ancient medicine. This great man combats this dangerous error, and shows that the foundation of medicine must be within the sphere of sense experience, that one must judge of foods and remedies by the relation in which they stand to nature and in accordance with the good or evil one receives from them, and not on the basis of imaginary hypotheses, as these innovators wished."

Before we leave this topic a final word needs to be said on the kind of hypotheses to which the writer of the tract on Ancient Medicine objects. He has, of course, no objection to an hypothesis in the modern sense, namely a summary statement of the conclusion to be drawn from the observation of phenomena, confirmed by the test of experiment. Such hypotheses are necessary to science, and such he himself constantly employs. What he does object to is a postulate, or general truth, regarded as self-evident and put forward not as subject to the confirmation of experience but as a basis for logical deduction. In the great debate between the reason and the senses the writer of the tract on Ancient Medicine is in the opposite camp to Parmenides. In the sphere of medicine at least he will not admit the validity of any argument however logical until its conclusions have been submitted to the test of experience. Antiquity soon lost sight of this truth and it remained in neglect until rediscovered in the dawn of modern science. Thus Roger Bacon in his *Opus Majus* writes : " There are two ways of knowing, by reasoning and experience. Reasoning draws conclusions and makes us do so, but it gives no feeling of certainty and does not remove doubt so that the mind may rest in the vision of truth, unless the truth be discovered by the path of experience."

So far as our story has gone up to the present it has been notable how little contribution to thought has been made by the mainland of Greece. About the middle of the fifth century, however, Athens assumed the lead in culture as she had already done in political power among the Greeks. The predominant part she had played in repelling the advance of Persia in the first decades of the century combined with her command of the sea gave her the undisputed leadership not only of the mainlanders but of the Asiatic Greeks. The fair temples which Pericles caused to be built out of the tribute of the allies became the external symbol of her power and taste, while the new drama of Æschylus and his successors established and maintained her intellectual prestige.

The drama, which admits of an audience of tens of thousands, was the natural art-form to be taken up and encouraged by a democratic city-state, just as the bard-sung epic had been appropriate to a régime of petty kings and feudal retainers, and as the personal lyric was the natural medium of expression of individualistic aristocrats. But the rich philosophic content of the Athenian tragedy is proof of the volume of external influences which, with the achievement of power, began to pour into the comparatively empty spiritual reservoir of the Attic capital. In comparison with Miletus, Ephesus, Croton, Elea, Athens was culturally still a backward town. Now, as she assumed the leadership of the Asiatic Greeks, an age of enlightenment set in, and the fortunate preservation of many of the literary masterpieces of the time enables us to appreciate the impact of the new ideas on a receptive society. There is no work from any period in any literature which more powerfully communicates the spiritual excitement of a society which stands both exalted and dismayed before the prospect of a revolution in thought than the *Prometheus* of

Æschylus. Prometheus, the giver of the arts of civilization to men, is the symbol of the conscious effort of mankind to take his destiny into his own hands and mould a fairer world. But the break with the past is not easy. The old instinctive allegiances still assert their claim. Is there not presumption in innovation? Is there not risk in displacing the old landmarks? Prometheus is made to suffer for his pride. For those who are interested in the history of thought, not merely as a record of achievements supposed to be won in an atmosphere of philosophic calm but as a ferment in an evolving society, this play is a document of the first importance.

That the new ideas encountered resistance in Athens is certain. As we have already said, the first philosopher to settle in Athens was Anaxagoras of Clazomenæ. Protected at first by the patronage of Pericles, he disseminated the natural philosophy of the Asiatic and Italian Greeks as brought up to date by himself, and his views are reflected in many a passage of the tragedian Euripides. In the end he provoked the animosity of the conservative Athenian democracy. In a famous passage Milton speaks of Athens as " native to famous wits or hospitable," but in fact the wits, whether native or foreign, were often made to feel that they had outstayed their welcome. Æschylus ended his days in Acragas, the birth-place of Empedocles. Euripides died in Macedon. And the astronomical theories of Anaxagoras were more than the Athenians, who worshipped the heavenly bodies as divine, were prepared to tolerate. Anaxagoras has the credit of being the first man to teach clearly that the moon shines by reflected light, and this knowledge enabled him to advance to a correct explanation of solar and lunar eclipses. These discoveries were connected with views on the nature of the sun and moon that seemed blasphemy against the gods Apollo and Diana with whom the sun and moon were identified. Anaxagoras was indicted for teaching the impious doctrine that the sun is a red-hot stone and the moon made of earth. He was banished, or to escape persecution fled, from Athens and ended his days in Lampsacus on the Dardanelles.

Another of the foreigners attracted at this time to the chief city of the Greek world was Hippocrates of Chios, a mathematician, not to be confused with his namesake, the doctor of Cos. The story goes that owing to the loss of his property he was reduced to the necessity of earning his bread by teaching and remained in Athens pursuing this profession for many years, probably from about 450 to 430 B.C. Necessity is the author of many books, and Hippocrates is distinguished as being the first to compose a text-book of geometry. His Elements prepared the way for Euclid's later masterpiece. He also made brilliant contributions to the problems of squaring the circle and doubling the cube, but we do not hear that he was banished. Physicists have often been in trouble with religious orthodoxies, mathematicians never. Mathematics and theology in so far as they can be regarded as deductive sciences independent of experience are natural allies. God is always a geometer.

With the rise of big commercial centres like Athens and, to choose an example from the west, Syracuse, teaching for money became common. Itinerant professors, known to history as Sophists, are a characteristic of this period. They moved from town to town, teaching mathematics, lecturing on medicine or astronomy, giving lessons in rhetoric and politics. Some of them made important contributions to knowledge. Hippias of Elis invented a curve for trisecting any angle, known afterwards as the quadratrix when its application to the problem of squaring the circle became clear. Other notable sophists were Gorgias of Leontini in Sicily, and Protagoras of Abdera, a fellow-townsman of Democritus, to both of whom we shall have occasion to refer again. They were the forerunners of our modern exponents of higher education. But they were subsequently held up to scorn by Plato, who had an independent income, for teaching for money, and their name has become a term of reproach.

Essentially the same type of man as the sophists,

and typical of the restless activity of his age, was the great historian Herodotus. Born about 485 B.C. in Halicarnassus he, like so many others, decided to move west on account of political unrest. He threw in his lot with the colonists who were bound for Thurii in Italy, the town for which Hippodamus of Miletus made the plans, and on his way thither in 445 B.C. he stopped in Athens and gave public readings from the historical work on which he was engaged. This work, in its final form, has survived the wreck of ancient Greek literature and constitutes the first important extant work of European history. It is remarkable for its scope. Just as Anaximander had attempted to map the civilized world, so Herodotus attempted to write its history. His main theme is the struggle between east and west, between Persia and Greece, culminating in the battles of Marathon and Salamis, but he begins with the ancient civilizations of the east, recording in his early books what he knew, or wished others to think he knew, of the rise of Egyptian and Babylonian civilization.

Herodotus was a product of Ionian intellectualism, and as such his spiritual debt to the *Iliad* is immense, and in the case of history the debt is naturally more direct and obvious. Like Homer he understands an historical movement as a complex of character and situation. Accordingly, the events of his story unfold themselves out of a background of national types and bear on them the impress of the characters of the chief actors. History of this type is secular and scientific in its central conception, and it is not surprising to find Herodotus explicitly state that he avoids religious controversy as much as he can. Among the Hippocratic writings there is one, *Airs, Waters, Places*, in which Ionian naturalism, as distinct from supernaturalism, is applied with calm assurance to the elucidation of human history on the grandest scale. Anaximander had suggested that man is a product of nature, a spontaneous generation of mother earth at a certain stage of her history. *Airs, Waters, Places* develops the theme, attempting to expound the effect of climate and geographical situation on health, and more generally the effect of climate upon character. The temper of Herodotus is the same. He is as confident that history is intelligible as Thales is that the cosmos can be understood.

A notable instance of his impregnation with the temper of Ionian science occurs in the course of his description of the physical features of Egypt. Geological speculation had made some headway among the Ionians. A hundred years before Herodotus, Xenophanes of Colophon, a philosophical poet, had noted that shells are found far inland and that in the quarries of Syracuse were *imprints*, as the Greeks called them, or as we say, fossils, of fish, seaweed, and so forth ; and from these observations he had drawn the conclusion that what was then dry land must once have been under water. Herodotus has a similar geological outlook on the formation of the earth's crust. Aware of the vast alluvial deposit carried down by the Nile and acquainted with the topography of the Nile Delta, he boldly champions the opinion that the whole of the Delta has been created by the alluvial deposit of the Nile. He is aware of similar phenomena in many parts of the Ægean area—he mentions five examples—and concludes by the suggestion that if the Nile were to reverse its course and flow into the Arabian Gulf it would fill it with soil in twenty, nay, in ten, thousand years. The significance of these casual calculations in tens of thousands of years, and of the conception of huge alterations in the structure of the globe as being the result of the slow operation of natural causes, will be the more apparent when we recollect how soon men lost the understanding of nature that is necessary to enable them to be made. " The poor world is almost six thousand years old," says Rosalind in *As You Like It*, voicing the tradition of an epoch in which natural science had long been submerged by revelation. And as has been already mentioned, Rosalind's view of the age of the world was still orthodox when Queen Victoria came to the throne of England. Such narrow conceptions as those in vogue in England under Elizabeth and Victoria would have seemed ignorant to Herodotus.

Even more remarkable as evidence of the penetration of the Greek mind in the fifth century by scientific conceptions is the history of the great rival of Herodotus, Thucydides. Openly contemptuous of importing the supernatural as a motive force into history, he deepened the Homeric and Herodotean conception of an historical event as a complex of character and circumstance by a profound realization of the importance of the economic factor in the evolution of society. In the light of these ideas he aspired to tell the story of the struggle for power between Athens and Sparta in such a way that the necessary connection between events should stand out clear as a lesson for posterity. He marks no abrupt break in historical tradition, but so clearly realized and so overmastering is his conception of history as being dominated by natural law that he is rightly called the first strictly scientific historian.

Such a conception of history is often criticized as if, like the atomic world of Democritus, it implied a denial of human will. This is not so. The causal connection which such a writer as Thucydides seeks to establish between events is not merely a mechanical causation, such as presided over the dance of the atoms in the system of Democritus, but it rests upon the belief that the human mind and character are products and parts of nature and, as such, proper objects of scientific research. The Hippocratic author of the tract *On Ancient Medicine* makes it abundantly clear that he regards the physical speculations of the natural philosophers as insufficient to explain the physiological activities of the human organism. Much less could Thucydides accept the view that the psychological element in society, the reason in man to which he directed his appeal, was reducible to physical laws. He was no fatalist. He took a tragic view of human destiny, but not a hopeless one. The purpose of his severe and noble book was to make later generations wiser than his own in the hope that they might avoid the disaster that had overtaken his.

When one pauses to look back on the achievement of the Greeks in the realm of science in the 200 years between Thales and the end of the fifth century, one understands how those who are careless of the unscientific implication of the term have come to call it a miracle. A sustained process of thought lasting over a few generations has transformed man's outlook on his world. Mathematics has advanced from a disorderly handful of practical rules, in which the idea of proof is only occasionally to be discerned, to an organized science in which the idea of building up the whole structure as a logical deduction from the smallest possible number of axioms and postulates is clearly envisaged.

Astronomy, as far as the recording of the apparent movements of the celestial bodies are concerned, has been a lesson learned from Babylonia, but the science has nevertheless completely changed its character. It has ceased to be merely an astronomy of apparent position and become an effort to determine the substance of which the heavenly bodies are made and to ascribe physical causes for their motions. The problem of the constitution of matter has been solved by an atomic theory which bears a recognizable likeness to the hypothesis on which modern physics and chemistry were based, and the origin of the cosmos has been explained in terms which resemble the systems of a Kant or a Laplace and bring the heavens under the reign of the same laws as are presumed to operate on earth. These brilliant speculations on the constitution of matter and the origin of the universe have been made by a combination of observation and reasoning which commands even more respect for the soundness of its method than for the fact that the views arrived at happen to have anticipated modern conclusions. Finally, in the medical school of Cos, a still further advance has been made; for there it has been distinctly taught that in a sphere where the material under investigation is accessible to the researcher no conclusions based on deduction from hypotheses are acceptable until submitted to the test of experience.

Inevitably also a science such as this has had its repercussions beyond the narrow circle of active scientists. It tends to become a leaven in society as knowledge is recognized to be an element in right conduct. The old supernaturalism threatens to be outmoded and to pass away. Zeus is no longer required to wield the thunderbolt. Poseidon and his nymphs become adornments for verse. Disease is no longer a visitation from an offended deity. That fiery stone, the sun, and the earthy moon shining by reflected light, can with difficulty be identified with Apollo and Diana. A general who regards eclipses as a supernatural warning is felt to be a pathetic example of superstitious weakness as well as a public danger. And the oracle-monger and the sooth-sayer have to fight for their position in society.

An anecdote from Plutarch will illustrate the temper of the time ; it tells of the influence of Anaxagoras on the Athens of Pericles. " Among other advantages," writes Plutarch, " that Pericles had of his association with Anaxagoras was this, that it appears that he was lifted by him above superstition, that feeling that is produced by amazement at what happens in regions above us. It affects those who are ignorant of the causes of such things and are obsessed by the idea of divine intervention and bewildered through their inexperience in this domain ; whereas the doctrines of natural philosophy remove such ignorance and inexperience and substitute for timorous and inflamed superstition that unshaken reverence which is attended by a good hope.

" A story is told that once on a time the head of a one-horned ram was brought to Pericles from his country-place, and that Lampon the seer, when he saw how the horn grew strong and solid from the middle of the forehead, declared that, whereas there were two powers in the city, that of Thucydides and that of Pericles, the mastery would finally devolve upon one man—the man to whom the sign had been given. Anaxagoras, however, had the skull cut in two, and showed that the brain had not filled out its position, but had drawn together to a point like an egg at that particular spot in the entire cavity where the root of the horn began. At that time, the story says, it was Anaxagoras who won the plaudits of the bystanders. But a little while after it was Lampon who was applauded, for Thucydides was overthrown and Pericles was entrusted with entire control of the interests of the people.

" Now there was nothing, in my opinion, to prevent both of them, the naturalist and the seer, from being right ; the one correctly divined the efficient cause, the other the final cause or purpose. It was the proper province of the one to observe why anything happens and how it comes to be what it is ; of the other to declare for what purpose any-thing happens and what it means.   And those who declare that the discovery of the efficient cause in any phenomenon does away with the meaning, do not perceive that they are doing away not only with divine portents but also with conventional signs such as the ringing of gongs, the language of fire-signals, and the shadows of pointers on sun-dials. Each of these has been made through deliberate adaptation to have a special significance.   However, perhaps this is matter for a separate treatise." (*Life of Plutarch*, Chap. 6.)

The reconciliation by Plutarch of the claims of scientific causality and providential government of the universe is an interesting one.   It has been reinvented every time that the controversy between the naturalists and the supernaturalists has arisen. But it is important to observe that Anaxagoras gave reasons for his view and that cumulative experience has tended to strengthen the claim to validity of his method of interpreting phenomena. The reverse has been the case with Lampon's view. No civilized government now maintains a state seer for interpreting the political significance of biological freaks.

CHAPTER VI

SOCRATES AND PLATO

THE ATTACK ON IONIAN SCIENCE

PLATO and his master Socrates are the only two great thinkers who were natives of Athens, but no two names stand higher in the history of thought. Their contribution was so important that it is customary to divide the whole series of Greek thinkers into pre-Socratic and post-Socratic, and Socratic in this connection is almost synonymous with Platonic. For Socrates wrote nothing and is mainly known to posterity through the series of dialogues in which Plato dramatized his personality and preserved his thoughts.

Whether the revolution in thought effected by Plato and Socrates was beneficial or not for science has been, and often still is, matter of debate. There are those for whom Plato is equally great as philosopher and as scientist. There are others who look upon the Platonic influence as so fatal to science that they cannot admit that the tree of knowledge has ever grown under the shadow of his philosophy. The truth lies between these views. Plato fought Ionian science with a life-long passionate hatred ; Platonism bequeathed to mediæval thought an outlook that was incompatible with the growth of positive science ; and when science revived in western Europe in the sixteenth and seventeenth centuries it looked back beyond Plato to the pre-Socratic thinkers who had shared its temper and blazed its trail. But it could not simply resume the old Ionian tradition without being guilty of a gross anachronism. There is substance in Plato's polemic that must be reckoned with, so much substance indeed that the achievement of his predecessors is not unfairly described as the dawn of Greek thought. Plato cast some ugly shadows, but he brought an illumination like the rising sun.

The tendencies for which Plato fought had long existed in Greek thought, but they were incarnated in the person of his master Socrates, perhaps the best known and the best loved son of Athens, and one of the strongest personalities in the history of our race. Socrates, who was born in 469 B.C., was in his youth a follower of the Ionian tradition of natural philosophy and an associate of Archelaus who carried on the work of Anaxagoras at Athens. It is very probable that he was himself the centre of a group of interested enquirers into the philosophy of nature and that he was widely read in the writings of the Asiatic and Italian schools. But he became dissatisfied with physical science owing to its neglect of the conscious and purposeful element in man which he identified with the soul. A man of indomitable courage and a born reformer, it did not seem to him that the active, enquiring, stubborn spirit that ruled in his breast, and at whose behest he was willing at any moment to lay down his life, could adequately be accounted for by any of the materialist philosophies, nor did he think that these philosophies offered any clear guidance to the individual or to society as to their proper path in life. Only one sentence in all that the physicists had written seemed to him worth a straw. Anaxagoras had begun his book on the *Nature of Things* with the statement: " In the beginning everything was in confusion, then Mind came and reduced them to order." In the whole of his subsequent discourse Anaxagoras made no further use of the concept of mind ; but to Socrates a new path of enquiry had been revealed. He took no further interest in the effort to interpret the phenomenal world as a sequence of mechanical causes and effects. For, he thought, if it is mind that arranges things, then everything must be arranged for the best, and the enquiry into the causes of things ought to be an enquiry into what is best. This was the new impulse that he gave to philosophy and science. He started it on the quest for evidences of intelligent design in the universe in contradistinction to the reign of mechanical law.

The new impulse had not, of course, come entirely unprepared. What Socrates had hoped of Anaxagoras was that if he told him the earth was flat or round, or in the centre of the universe, he should also make clear that it was best that it should be as it was. We have met something like this before among the Pythagoreans who had decided that the earth was round because a sphere is the perfect figure. And there can be no doubt that in the doctrine of mind which he was now introducing, and to which Plato was to give such remarkable extension, Socrates was carrying on the Pythagorean tradition. Since this doctrine of mind or soul is what is most characteristic in the teaching of Socrates we must endeavour to fix more precisely the degree of its originality.

In the Homeric poems we find the soul represented as surviving the death of the body, but the soul is looked upon as a sort of shadowy duplicate of the real man who has died. Ionian science inherited this point of view. For it the soul was a material substance like the body, only made of finer stuff. When any account was attempted of the phenomena of mind and consciousness it was in terms of physical intermixture, Empedocles opining that the four elements in us recognized the four elements in the outside world, Democritus that streams of atoms carried into us through our sense organs images of the outside world. The atomists, it is true, made penetrating observations on the nature of sensation and the sensible qualities of things, going so far even as to maintain that such qualities as colour and heat and sound do not exist apart from the percipient but are effects produced on the organs of sense by the motions of atoms in the void. But this is still only physics. It does not even mark a beginning of psychology. The question of what consciousness is, of what perception is, cannot be said to have been raised until it is clearly distinguished from physical contact and intermixture.

A notion of the soul as immortal and somehow of different nature from the body has been presented to us in the teachings of Heraclitus and Pythagoras, both of whom had the idea of the soul's being imprisoned in the body and in danger of contamination by it. Heraclitus who regarded the soul as fire had advocated temperance on the ground that moisture quenches the sacred fire. " A dry soul," he said, " is best." And the Pythagoreans had taught the doctrine of metempsychosis or transmigration of souls, coupled with the idea of a happier rebirth for a soul that has been faithful in well-doing.

It seems clear also that in the more elevated circles of the Pythagorean brotherhood an interesting development had occurred. In the Greek world of the sixth and fifth centuries there were many mystery cults in which rites of purification were undergone by the initiates in order to fit the soul for the life after death. In these cults the emotional element seems to have predominated. But the Pythagoreans conceived the idea that *knowledge* is a purification, and the pursuit of mathematics in their brotherhood partook of the nature of an initiation of the soul into the life eternal. We have seen that at first their mathematics were of a half-material kind. Numbers and things were not clearly distinguished from one another. But at least by the time of the emergence of the atomic theory in the middle of the fifth century the distinction between mathematics and physics had become clear, and mathematics had emerged as an abstract science differing from physics as being independent of sense-perception. An opposition developed between the exponents of the two types of knowledge. Physical speculation seemed to be concerned with the transient world of phenomena, the flux of Heraclitus, which Parmenides had declared to be a mere illusion of the senses ; but mathematics, having as its subject-matter relations of space and number, was a science of the eternal and the changeless, not dependent on the treacherous evidence of the senses, but apprehended by reason or the soul alone. Mathematics was an initiation into the mind of God.

Socrates, if we may trust the Platonic account

of him, found in the nature of mathematical knowledge an argument for the immortality of the soul.
The objects of mathematical knowledge are not
derived from sense. The triangles and circles
that we find in nature merely serve as reminders or
visible counterparts of the ideal figures with which
the geometer operates. There is no such thing in
nature as a perfect circle, a perfect equilateral
triangle, or truly parallel lines. The ideal figures
of mathematics given in their definitions, and the
truths which mathematics deduces from these
definitions, are independent of experience. Once
we have been made to understand the properties of
the circle our conviction of their truth is not confirmed by increased acquaintance with the imperfect
circles existing in nature. Mathematical knowledge is independent of experience and is the
standard by which we judge experience. Its truth
is absolute, eternal, and changeless, and the souls
which know such truths (and Socrates undertook
to prove that the knowledge of them slumbers in
every soul) must have acquired the knowledge in
some other world than this, a world in which the
soul was in direct contact with the eternal verities
of which this world can show only fleeting and
imperfect imitations.

This mathematical doctrine of an independent
world of supra-sensible reality was given a remarkable extension to the domain of ethics by Socrates.

This was the matter that lay closest to his heart. The corruption of public and private morals in Greece by the course of the Peloponnesian War has been painted in the darkest colours by the historian Thucydides. It is typical of the rational character of Greek civilization that Socrates did not at this juncture assume the rôle of a spokesman of God in quite the same way as the Hebrew prophets were wont to do ; he endeavoured rather to create a science of ethics. His task was hampered by the sceptical and relativist outlook on philosophy and morals that had become common among intellectuals. A saying of the sophist Protagoras has been widely accepted as typical of the spirit of the age. " Man is the measure of all things," was his dictum, and it is quite likely that he meant thereby to deny the existence of absolute standards of behaviour. Another sophist, Gorgias, put in the forefront of his teaching three disturbing propositions : " There is no truth ; if there were, it could not be known ; if known, it could not be communicated." To this scepticism it was the ambition of Socrates to put a stop.

His method was to attempt to extend to ethical concepts the clearness and certainty that attach to the concepts of pure mathematics. Just as he was convinced that a knowledge of the eternal truths of mathematics slumbers in every soul, and that the process of teaching mathematics was not a bringing in of something from outside but an awakening of knowledge in the soul, so he thought that the ideal forms of virtue were part of the knowledge every man brings with him into the world. Perfect justice, perfect truth, perfect beauty, like perfect circles, do not exist in nature. Knowledge of them therefore cannot be derived from experience, but is part of the heritage the soul brings with it from its pre-existence in an immaterial world. Submerged in the body the soul has difficulty in recapturing the vision of the good. But a training in mathematics, by turning it from things of sense to pure forms, can fit it for philosophy, that is to say for the quest for the knowledge of absolute virtue.

Socrates also attempted, with less conviction and with less success, to extend the doctrine of forms to the whole of nature. There is much talk in the Socratic dialogues of the Idea or Form of man, of the horse, even of tables and chairs. The notion is that the whole visible world is a copy or imitation of an intelligible world of pure forms which is the true reality. This is the knowledge of which the soul ought to lay hold, the knowledge of the supra-sensible realities of its own eternal home. To stoop to the confused notions that can be acquired of the perpetually changing physical world through the treacherous medium of our bodily organs of sense is but to defile the soul. "If we are ever to know anything absolutely," Plato represents Socrates as saying in the *Phædo*, " we must be free from the body and must behold the actual realities with the eye of the soul alone. And then, as our argument shows, when we are dead we are likely to possess the wisdom we desire and profess to be enamoured of, but not while we live . . . And while we live we shall be nearest to knowledge when we avoid, so far as possible, intercourse and communion with the body, except what is absolutely necessary, and are not infected by its nature, but keep ourselves pure from it until God himself sets us free."

Filled with enthusiasm for his new conceptions of absolute virtue and absolute knowledge and for the doctrine of the immortality of the soul, Socrates became a self-constituted, or as his own deep instinct bade him say, a God-appointed, missionary to his fellow-citizens, urging them in season and out to pay heed to the salvation of their souls and to let no other consideration weigh with them in comparison with this. When in his seventieth year a tyrannical government which hated his independent temper brought him to trial on a trumped-up charge, he took the opportunity of his impending martyrdom to commend for the last time to the Athenian populace his gospel of the immortality of the soul ; and his last hours before he drank the prescribed draft of hemlock were spent in expounding his deeper thoughts on this subject to the intimate circle of his habitual friends and associates. These scenes have been immortalized by Plato in pages that are acclaimed to this day as among the most sublime achievements of literature. But the believer in the pre-Socratic tradition of physical science cannot regard the new orientation given to Greek thought by Socrates without regret. To Socrates physical science was a complete irrelevance, or rather a positive ill ; and the moving record of his trial, last days, and death, read and re-read by generation after generation of youthful scholars, is unfortunately inextricably inwoven with a polemic against the body as a source of contamination, the senses as a snare, and the world as a vain and fleeting show.

Socrates had performed an essential service in protesting against the pretensions of the physical science of his day to be a complete account of reality. His assertion that the soul in man is an active and not merely a passive principle, and his rejection of the efforts to explain its activity on the basis of the physical interaction of material particles, prepared the way for the creation of a genuine psychology. But his complete revolt from physical enquiry was one-sided and reactionary and had evil results. From now on mathematics, ethics, and theology become inextricably blended as *a priori* sciences independent of experience giving us the only truth we know and sharply opposed to physical science, which is condemned as materialistic, atheistic, and sunk in the " dregs of the sensuous world." It has often been claimed that the advent of Christianity was the signal for the downfall of ancient science. But the contempt for the physical world which was one of the main reasons for the death of science had already found full expression in the philosophy of Socrates. And, as we shall see, the temper that would persecute a Galileo or a Bruno for a physical theory was about to be born in the mind of Socrates' chief disciple. In this, as in so much else, Plato was the originator of Christian theory and practice.

Plato, born in 428 B.C., was approaching his thirtieth year when his master Socrates was the victim of judicial murder by the restored democracy of Athens. Abandoning in disgust his ambition for a political career he devoted himself to travel and study and the composition of a series of dialogues in honour of the life and teaching of his dead master. Returning to Athens in his forties he founded an institution, the Academy, for higher learning, with the double ambition of promoting true science and training up a generation of public spirited and instructed men who might rescue Greece from political anarchy. Probably for about twenty years his energies were wholly absorbed by his duties as director of studies. But in the last fifteen or twenty years of his life, which did not terminate till he was over eighty, he produced a second series of dialogues, less charming as dramatic compositions than his earlier work but of greater philosophical import.

Plato was a voluminous writer and his works have come down to us in their entirety ; it might therefore seem that we are in a peculiarly favourable situation for knowing exactly what he thought on a vast variety of subjects. This is not so. In his first period of composition he seems to have been mainly concerned to perpetuate the thoughts of his master Socrates, and it is not clear how much he shared them. Of his teaching in the Academy he left no record. And in his later writings he again adopted the dialogue form, making Socrates himself, or Eleatic or Pythagorean philosophers, the chief speakers. No complete or consistent system can be gathered from these dialogues, early or late, nor is it to be expected that it should. Plato explicitly states that his system, so far as he had one, could not be and *a fortiori* had not been committed to writing. Philosophy was not for him a series of propositions that could be grasped by a clever student, but a view of reality personal to each student. Reality, of course, was the same for all, but the apprehension of it was more intimate than logic and depended in the last analysis on individual intuition.

The problem of the essential nature of Platonism

is, therefore, a very difficult one. Our concern, however, is not with it, but simply with the bearing of his published writings on the question of the possibility of a true science of nature. The question is one that comes up again and again for discussion in the dialogues and results of the greatest importance are achieved. Whether the various solutions propounded represent the ultimate conclusions of Plato or not is another matter. Any opinion which it seems the main purpose of a Platonic dialogue to advance, no matter who the speaker may be, is for our limited purpose Platonic.

That the Academy at first was wholly under the influence of the Socratic revolt against physical science seems a matter beyond question. In the *Republic*, the greatest dialogue of Plato's earlier period, which was composed about the time of the foundation of the Academy and clearly covers the range of subjects that were to be studied there, Socrates is made to express an attitude to astronomical studies that has become notorious. "The starry heaven which we behold," he says, "is wrought upon a visible ground, and therefore, although the fairest and most perfect of visible things, must necessarily be deemed inferior far to the true motions of absolute swiftness and absolute slowness. . . . These are to be apprehended by reason and intelligence, but not by sight. . . . The spangled heavens should be used as a pattern and with a view to that higher knowledge. . . . But a true astronomer will never imagine that the proportions of night and day, or of both to the month, or of the month to the year, or of the stars to these and to one another, and any other things that are material and visible can also be eternal and subject to no deviation—that would be absurd ; and it is equally absurd to take so much pains in investigating their exact truth. . . . In astronomy, as in geometry, we should employ problems, and let the heavens alone if we would approach the subject in the right way and so make the natural gift of reason to be of any use." (*Republic*, VII, 529.)

This, of course, is exactly in line with the previous recommendation in the *Phædo* " to be free from the body and behold the actual realities with the eye of the soul alone." Since the attitude here expressed is almost certainly that of the Academy at its foundation, it is of first importance for determining the position of Platonism in the history of science. In the modern effort to interpret nature mathematics has played an essential part. " When you can measure what you are speaking about," wrote Lord Kelvin, " you know something about it ; when you cannot measure it, your knowledge is of a meagre and unsatisfactory kind." Plato, by the direction he gave to the studies in the Academy did much to promote the progress of mathematics. But it is hardly correct to argue from this, as is so often done, that Plato therefore laid the foundation of modern scientific method. The ambition of Plato was not to interpret nature by the aid of mathematics but to substitute mathematics for physics. In the passage just quoted he recommends the abandonment of astronomy as an observational science and the substitution for it of a theoretical astronomy as a branch of pure mathematics. It is a mischievous inroad of philosophy on the domain of science. A defence of ancient astronomy on the lines of the Hippocratic defence of ancient medicine was badly needed.

The reactionary nature of the Platonic attitude is still more clear when we take into account the historical context. When Plato deprecates the pains that are taken to investigate the exact relations of the lengths of day and night, of both to the month and of the month to the year, and of the courses of the other stars, he is not making an academic point but attacking one of the useful activities of his day. The more precise determination of the relation to one another of the natural divisions of time with a view to the construction of a more accurate calendar was one of the age-old problems in the application of science to everyday needs. The Babylonians and the Egyptians had both felt the difficulty and met it, as we have seen, with an impressive measure of success; and now for a generation the problem had been pressingly before the instructed public of Athens. In the words of a later Greek astronomer, Geminus of Rhodes, who wrote about 70 B.C., " the ancients had before them the problem of reckoning the months by the moon but the years by the sun." As we now know the solar year is $365\frac{1}{4}$ days approximately, while the month is approximately $29\frac{1}{2}$ days. Accordingly, if we reckon 12 months to the year the calendar falls short of the sun by some 11 days each year ($29\frac{1}{2} \times 12 = 354$). Efforts were accordingly made to determine a cycle of years in which the lunar year and the solar year would exactly correspond. This led to an ever more precise determination of the exact length in days of the month and year, when the calendar, at each successive adjustment, was still found to be not completely in accord with the observed phenomena of nature.

The Greeks as early as the eighth century had borrowed from the Babylonians an eight-year cycle. In eight years the shortage of 11 days a year in the calendar amounts to 88 days. This is almost the equivalent of 3 months ; and the insertion of 3 intercalary months at intervals throughout an eight-year cycle, together with the adoption of a month that varied between 29 and 30 days, produced a fairly close agreement of the lunar with the solar year. A sixteen-year cycle was later adopted. Then in the year 433 B.C., five years before the birth of Plato, an astronomer Meton introduced a nineteen-year cycle into Athens. This was the state of the calendar at Athens when Plato wrote the *Republic*, and his advice, if it had been heeded, would have precluded the invention of the seventy-six-year cycle of Callippus, who came to Athens about 334 B.C., and of the 304-year cycle that was made by Hipparchus about 125 B.C.

The cycle of Hipparchus depended upon an estimate of the length of the tropic year of 365 days, 5 hours, 55 minutes, and 12 seconds, which is about $6\frac{1}{2}$ minutes too much by our reckoning, and a length of the mean lunar month of 29·530585 days, which is correct to four places of decimals, and is less than a second out. Such were the triumphs of science from which Plato would have deterred Greek astronomers. If he had succeeded the Greeks would not even have shared with the Babylonians the credit of the precise determination of the calendar, so necessary to an elaborately organized civilization. For in our admiration of the results achieved by Hipparchus we must not overlook the fact that the Babylonian astronomer Naburiannu, who flourished about 500 B.C., had already given a determination of the length of the mean lunar month which is correct in our notation to three places of decimals, and that another Babylonian astronomer, Kidinnu, about 380 B.C., or some 250 years before Hipparchus and just about the time when Plato was busy with the *Republic*, had given an even more accurate figure than that of Hipparchus himself. What is more, Hipparchus was familiar with the Babylonian results. In the measurement of time the Greeks were still the pupils of the Babylonians. It was only in physical astronomy they surpassed them.

If we return now from the achievements of the practical astronomers in calendar making to the consideration of the scientific outlook of the Academy, it is not difficult to produce evidence that the Socratic conception of pure mathematics as the science of reality and as the necessary foundation of a sound grasp of moral principles continued to dominate the activities of the institution. This is illustrated by a tragi-comic interlude in Plato's career. When he was already sixty years of age and had been for twenty years director of studies in the Academy he was called upon to put into practice his principle that the true foundation for states-manship is the study of philosophy. Summoned to Syracuse to act as advisor to the young tyrant Dionysius II, who had just acceded to the throne and was anxious to put his city, then the greatest in the Mediterranean world, under the guidance of the world's greatest philosopher, Plato, in considera-tion of the loyalty he owed to philosophy, obeyed. On his arrival he at once began to instruct the prince and his courtiers in geometry. There was no other path to wisdom and virtue. The royal court took kindly at first to the new discipline, and, as they enthusiastically constructed their figures on the sanded floors, the whole palace, as Plutarch puts it, became one whirl of dust. But it is small wonder that before long Plato was swept out of court along with other encumbrances to a more practical policy.

It was after his return from Syracuse that Plato began a new series of compositions. These are characterized on their political side by an abandonment of the intransigeant idealism of the *Republic*, and on their scientific side by a parallel effort to assess anew the claim of the senses to be avenues to a knowledge of reality. The first dialogue in which the new orientation is revealed is the *Theætetus*, the composition of which is generally recognized to have coincided roughly with the period of the Sicilian essay in practical politics. Here the abandonment of the Socratic advice to be done with the body and trust the eye of the soul alone is unmistakable, and the resumption of the debate on the respective contributions of reason and sensation to knowledge yields results of classical importance. Plato is now prepared to consider the position that the data of sensation are the materials of knowledge ; but he insists that sensation is not in itself knowledge. For the first time in history he clearly distinguishes between sense-perception and thought and teaches that knowledge is the result of the action of the latter on the former. " The simple sensations which reach the soul through the body are given at birth to men and animals by nature, but their reflections on these and on their relations to being and use are slowly and hardly gained, if they are ever gained, by education and long experience."

In the passage from which we have just quoted Plato propounds for the first time the idea that is axiomatic in later thought, that the sensory faculties are organs by which the mind apprehends external nature. " We do not see with the eyes but through them. We do not hear with the ears but through them. Nor could any one sense itself distinguish between its own activity and that of another sense. There must be something connected with both— call it soul or anything else you like—*with* which we truly perceive all that is conveyed to us through the sensory faculties. It is the soul, or *psyche*, that makes us aware that we perceive and that distinguishes the data of one sense-organ from those of another." Such, in condensed form, is the argument by which Plato lays down the foundations of a new science, psychology, the science of the soul.

Nor does this exhaust his discussion of the matter. We have other psychic activities even less directly dependent on sense-stimulation than the activity of distinguishing between the information supplied by the various sensory faculties or acting as a repository for their data. Such activities are memory, expectation, imagination, not to speak of the higher operations of the mind by which we apprehend such truths as those of mathematics, or by which we reason about the first principles of knowledge and existence. It is not any of the senses but the soul that apprehends such concepts as Being and not-Being, the Like and the Unlike, the One and the Many, the Beautiful and the Ugly, the Good and the Bad.

Here, then, as we have said, is the material of a new science, the science of mind or soul, the central concept of which is just as legitimate a deduction from experience as the concept of the atom. The distinction between mind and matter is the pivot on which the mature philosophy of Plato turns; and it is a contribution of such moment as fully to entitle Plato to his pivotal place in the history of thought.

In a later dialogue, the *Sophist*, the immateriality of the soul is strongly emphasized and it is shown how the materialists may be forced to admit its existence or adopt a very unreasonable attitude. The materialists are to be asked whether they do not admit the existence of such a thing as a soul, and that some souls are wise and good, others foolish and bad. If they say Yes, as they must, they are to be asked whether this does not imply that wisdom and the other virtues are something, and whether they are anything that can be seen and handled. Even if they try to save themselves by saying that the soul is a kind of body, it is hardly to be supposed that they will venture to say that wisdom is a kind of body, nor yet to say that it is nothing at all, though a thorough-going materialist would have to take this alternative. But with any who admit that a thing can *be* without being a body, the point has been gained.

In the *Timæus*, the one dialogue which Plato devoted expressly to physics, the doctrine of soul receives a new and more dubious extension. In this strange work, which contains much pure fantasy inwoven with an account of the creation of the world by a geometer god of Pythagorean derivation, it seems to be implied that the immaterial soul is the origin of the material world. God, it is said, who is a soul, made the sensible world on the pattern of an intelligible model. But whether he is supposed to have made the sensible world out of nothing or merely to have imposed order on a pre-existent chaos is not clear. To the subsequent influence of this puzzling book we shall refer in a later chapter.

It is in the *Laws*, his last, his longest, and his greatest work, that Plato gives definitive expression to his hatred of the Ionian school of physical science, and endeavours to set up in opposition to them a spiritual view of the constitution of the universe. The natural philosophers, he tells us,

say that earth, air, fire and water all exist by nature
and by chance, and none of them by design. They
say that the bodies that come next in order of evolution, viz. the earth, the sun, the moon, and the
stars, have been created by means of these absolutely
inanimate existences. The elements are severally
moved by chance, that is to say by some inherent
force, according to certain affinities among them—
affinities of hot and cold, or dry and moist, or
soft and hard. In this way the whole heaven has
been created, and all that is in the heaven, as well as
animals and plants ; all the seasons too come from
these elements, not by the action of mind, according
to these philosophers, or from God, or by design,
but by nature and chance only. Design, they say,
sprang up afterwards and out of these. It is mortal
and of mortal birth. Medicine, husbandry, and
legislation are examples of these later-born arts,
which are designed to supplement nature. The
gods likewise, these philosophers teach, are products not of nature but of art, being constituted by
the laws of the different states in which they are
worshipped. Thus religion is a product of art,
and so also is morality. The principles of justice
have no existence in nature, but are a mere convention. Thus does Plato define the outlook of
the natural philosophers of his day. And his
statement, making allowance for the great growth
of positive knowledge, would be accepted by many

scientists at the present day as a fair description of their outlook and aims. It is still an illuminating summary of the materialistic, or rather mechanistic, position.

In opposition to this he then expounds his own position. The physicists teach that earth, air, fire, and water are the first elements of all things ; that these constitute nature ; and that the soul is afterwards formed out of these. In other words, the order of evolution is inanimate matter first and life and mind afterwards. Plato's view is the reverse of this. For him the soul is the first of things ; it is before all bodies, and is the chief author of their changes and transpositions. The things of the soul or mind come before the things of the body. That is to say, thought, mind, art, law are prior to the hard, the soft, the heavy, the light. First comes design, or mind, and after it come nature and the works of nature. What is called nature is under the government of design and mind. It might be better to say that the term nature has been wrongly applied. The physicists in their use of the term mean to say that the four material elements constitute the first creative power. But if soul, and not matter, can be shown to be the first creative power, then in the truest sense and beyond all other things soul may be said to exist by nature.

That this is so Plato attempts to demonstrate by an analysis of motion. Animate things are to be distinguished from inanimate by this that they contain in themselves a principle of motion. Every living thing is self-moved; inanimate things can only move in so far as motion is imparted to them from outside. The ultimate source of motion in the universe must be some self-moved thing. This self-moved existence is the life-principle or soul, which Plato finally defines as a " motion that moves itself."

The argument from the analysis of motion is not convincing. Metaphysicians have delighted in the subtlety of the thought and warned materialists not to suppose Plato to have called the soul a *thing* that moves itself, and which therefore might be separated from its motion, but a *motion* that moves itself. But it is surely open for the materialist to retort that though a thing that moves itself might set other things in motion, a motion that moves itself, whatever such a concept may be supposed to import, could not set any material thing in motion. Plato's soul remains a metaphysical abstraction, and his attempt to include it under the concept of nature is a failure. Here as elsewhere the capital defect of his system, the absence of any bridge between the world of matter and the world of mind, is plainly apparent. But his general argument for the priority of design in the universe is on a different footing. The issue, as he has defined it, is still the greatest issue in philosophy and science, and his statement of the problem is so profound and so clear that one can still return to it after reading modern contributions to the debate with profit and pleasure. Those modern thinkers who believe in the independent existence of a physical universe, and regard life and consciousness as having somehow emerged out of it, are in the position of the old Ionian physicists. Those who tell us, on the other hand, that it is wrong to regard life as something that has intruded into or emerged out of a physical universe, but that the physical universe, as we know it, is merely the interpretation of the experience of a living thing, and that therefore life comes first and inanimate nature afterwards, are in the line of Platonic thought; they support with Plato the priority of soul.

In conclusion it must be said that the main purpose of Plato's life was to found a religion rather than to advance science. To him, of course, the two were theoretically the same; and he took steps to ensure their practical identity by recommending the institution of houses of correction for heretics. Under the system of government envisaged in the *Laws* citizens were required to hold in honour not only the Olympian gods and the gods of the underworld, but demons, heroes, and private and ancestral gods, and they were taught that to hold converse with these gods by means of prayers and offerings and every kind of service is the noblest of things and the most conducive to a happy life. Those who resisted these beliefs and rebelled against these exercises because of their sympathy with the theories of the physicists were to be put in concentration camps and instructed by state officials in the prescribed view of truth. If they proved obstinate they were to be subjected to hardships and further indignities, and, if these measures proved insufficient, to be put to death. In his advocacy of persecution for opinion Plato was also a pioneer, anticipating the discoveries of later centuries. To the historian of science it must ever remain a scandal that the brilliant intellect that founded a theory of mind as a rational deduction from observation and experience, and who thus immeasurably enlarged the domain of science, should have blended his arguments with the advocacy of exploded superstitions whose authority he sought to impose by the methods of the Inquisition.

CHAPTER VII

ARISTOTLE

THE RESTORATION OF THE IONIAN TRADITION

To the historian of science Aristotle appears as the genius who adapted the Platonic philosophy of mind to the requirements of positive research.

Born in 384 B.C. at Stagira in Macedon he was brought up at the court of the Macedonian king where his father was royal physician. It is probable, since this was the practice in medical families at the time, that he was apprenticed as a boy to the medical art and got some training in dissection. If so the germ of research was implanted in him in his early years. But obviously the career of a doctor did not satisfy him. When he was seventeen or eighteen years of age he came to Athens and entered as a student at the Academy. At this date the *Republic* and the *Phædo* were already classics. Aristotle must have been attracted to the idea of acquiring knowledge without the aid of the body by means of the eye of the soul alone. He remained a contented student of the Academy for some twenty years, that is until the death of Plato.

It must be born in mind, however, that Aristotle's long pupillage coincided with the new orientation towards the problem of reason and sensation that is pre-supposed by the *Theætetus*. The Socratic theory of Ideas would, of course, be still current in the Academy, but the new direction of thought would be to substitute for it the later Platonic doctrine of matter and mind. With the arrival of Aristotle as a youth at Athens coincided also that of the great geographer and astronomer, Eudoxus of Cnidus, then a man of some forty years of age. He was director of a school of research on Ionian lines at Cyzicus ; this school he transferred to Athens and became himself for a time a member of the Academy. In spite of the fact that he held views on the nature of the Good that were obnoxious to Plato (Eudoxus identified the Good with pleasure), his character made a lasting and favourable impression on Aristotle who records in his *Ethics* that he was a man of outstanding moral excellence. It is a very probable opinion that Eudoxus was the channel for the introduction of Ionian science into the Academy. His presence and influence would impose on Plato the necessity of taking the claims of physical science more seriously into consideration.

However that may be, it is certain that the career of Aristotle represents a progress from Socratic idealism, through the later Platonic recognition of the importance of sense knowledge, to a complete restoration of the practice of research as it had culminated among the Ionian Greeks in the Hippocratic school. These three phases of his intellectual life are clearly marked by the nature of his written works. In the first period, when he was still a member of the Academy, he wrote Socratic dialogues in the manner of Plato. In his second period, after the death of Plato, he abandoned the Academy because of its tendency, under Plato's successor and nephew Speusippus, " to turn philosophy into mathematics." But regarding himself as the true heir of Platonism he addressed himself to the problem defined by Plato but left unsolved, namely, how the mind acquires an understanding of matter. In his third and last period, when Aristotle was director of a school of his own at Athens, the Lyceum, he devoted himself with passion and energy to positive research, chiefly in the realm of biology. This period lasted twelve years, from 335 to 323 B.C. Roughly the first fifty years of Aristotle's life were devoted to getting his philosophical position clear. In the last twelve he reaped an unprecedented harvest.

The realization that the positive researches of Aristotle were mainly concentrated into the last period of his life is of recent date. His writings have come down to us in great bulk, but not prepared by him for publication. In his various treatises, which were kept in a fluid state during his life and continually revised and added to as his views developed and his researches progressed, some portions are early and some late, some Platonic and metaphysical in outlook, some wholly informed with the temper of positive research. Only the patience and insight of a contemporary scholar have forced the recognition that the earlier writings are those that are more Platonic in tone, while the positive researches are to be ascribed to the last period of his life.

Great damage has been done to our understanding of Aristotle by the failure hitherto to appreciate the development of his thought, and by the glib acceptance of him as an encyclopædic writer all of whose productions, no matter on what topic, are equally mature. Aristotle wrote on physics, on logic, on metaphysics, on psychology and biology, to mention only the topics relative to our discussion. So long as any of these various writings could be simply referred to as Aristotle it was impossible to make any consistent picture to one's mind of what, in the case of a great scientist, it is most important to understand, his method. But once it is realized that in his physical treatises he is largely under the domination of the Socratic outlook as set forth in the *Phædo* and the *Republic* ; that his logic and metaphysics represent his effort to adapt the later Platonism to the requirements of a science of nature ; and that his biological treatises contain the results of his later positive researches, his mental history becomes clear. It also becomes of unmatched significance for the student of the history of scientific method. Our purpose in this chapter is briefly to illustrate the three stages in his career.

In his treatise *On the Heavens*, which belongs to the period when the Socratic teleology of the *Phædo* determined his point of view, Aristotle is concerned to explain why the structure of the universe must be such as it is. The defect of this treatise is not that his ignorance of the structure of the universe seems to us gross; ignorance is always a relative term. It is that he is incurious, that he shows insufficient understanding of the necessity for observation and experiment, and that he satisfies his mind with verbal explanations drawn from the Pythagorean amalgam of ethics with mathematics. We are told that the heaven is a sphere, *because* a sphere is the perfect figure, and that it rotates in a circle, *because* only circular motion, which has no beginning and no end, can be eternal. When the supposed daily rotation of the heavens has been thus accounted for, we are next told that, as the centre of a rotating body is at rest, *therefore* the earth is at rest in the centre of the universe. Obviously the wreck of the brilliant Ionian cosmology is complete; and equally obviously the wreck is the result of approaching

physical enquiry on the basis of a strict adherence
to the theological guesswork of Socratic science
combined with the theological astronomy of Plato's
later years.

From Thales on the Ionians had based their
explanations of " the things above," as they called
celestial phenomena, on the assumption that the
stuff of the heavens was the same as the stuff of
earth ; for them the heaven, like the earth, was
subject to change, decay, and death. But for Plato
the stars were divine beings, changeless and eternal,
the regularity of whose motions was due to their
being instinct with reason or soul. This is the
point of view of Aristotle in his physical treatises ;
let us follow him as he continues to construct his
pattern of the universe with the eye of the soul
alone.

It is the nature of earth, he says, to be cold and
to move downwards. The earthy element must,
therefore, be balanced by its contrary, fire, the
nature of which is to be hot and move upwards,
for heat is prior to coldness, being the positive
quality of which coldness is the negation. But
these two bodies require two other bodies to mediate
between them. The reason for this is borrowed
from the Pythagorean number-physics of Plato's
Timæus. Solid bodies, having three dimensions,
correspond to cube numbers. Cubes require two
means to unite them, e.g. 1 and 8 are united by the

means 2 and 4 ($1 : 2 : : 4 : 8$). *Therefore* there must also exist two other elements, water and air. The necessity for the existence of the four elements of Empedocles has now been mathematically derived.

The method of this travesty of science is clear. It consists in borrowing the conclusions of earlier thinkers, dressing them up in a veil of verbal logic, and passing them off as proofs of the design in the universe which is apprehended by the eye of the soul. It is not to be denied that there is a great amount of material from the earlier thinkers preserved in the physical treatises of Aristotle; nor can it be denied that he shows great ingenuity in his endeavour to exhibit the logical necessity of what he supposed to be the facts. The trouble is that progress is checked by the emphasis that is put on logic to the detriment of observation. Long ago Parmenides had insisted that only the logical can exist. His exact words are : " It is the same thing that can be thought and can be." But man finds his powers of thought increase greatly with his experience. Science is on the whole in agreement with the popular point of view in this matter, that seeing is believing. But at this stage of his career Aristotle was rather inclined to the contrary view that whatever is logical must exist.

In this way, by the aid of the eye of the soul alone, he discovered a fifth element. Its existence was proved by an argument from motion. There are, according to Aristotle, two main kinds of motion, namely, up and down, and circular. It is the property of the four Empedoclean elements to move in the former fashion ; fire and air move up, they have the property of levity ; earth and water move down, having the property of gravity. What, then, naturally moves in a circle ? According to Aristotle it was a fifth element, ether. This element, moving with circular motion which is eternal, is itself eternal and not, like the other elements, subject to change. Of it the heavenly bodies are made. Thus by irresistible logic he had provided a physical basis for the Platonic belief in the eternity of the stars. Out of these five elements he proceeded to construct his universe.

In this design also he was greatly assisted by the mathematical achievements of the Academy. To the members of the Academy, who saw in the orderly movements of the stars a manifestation of the divine mind, one class of heavenly bodies was an occasion of offence. This was the planets, or vagabonds as their name means in Greek, whose disorderly progressions and retrogressions constituted an exception to the regularity otherwise observable in the heavens. Plato, therefore, set before the Academy the problem of devising a mathematical scheme of regular motions for the planets which would account for the confused appearance their motions present to our observation. This he called by the famous phrase, " saving the appearances." The problem received a remarkable solution by the mathematician Eudoxus, who analysed the apparent paths of the planets into the resultants of some thirty odd circular rotatory movements. The number of these spheres of revolution was considerably increased by a mathematician, Callippus, who belonged to Aristotle's generation and whose elaboration of the scheme of Eudoxus is proof of the esteem in which it continued to be held.

It was this emended scheme that Aristotle adopted as the plan of his universe. From a mathematical hypothesis he transformed it into a mechanical construction, this step being facilitated by his doctrine of the fifth element of which he imagined the celestial spheres to be made. His idea was that the earth consists of four concentric spheres made of earth, water, air and fire, proceeding in order from the centre outwards. These four spheres are not clearly defined from one another, as among these terrestrial elements there is a constant process of interchange. Outside these four terrestrial spheres were ranged in concentric circles fifty-five celestial spheres, revolving about the stationary earth and carrying with them in their revolutions the various heavenly bodies. The outermost sphere was that of the fixed stars. These fifty-five celestial spheres together with the heavenly bodies were made of the fifth element of ether and were changeless and indestructible. The lowest of the fifty-five spheres was that of the moon ; and it was only below the moon, in the sublunary spheres, that change and decay were to be found. Thus the religion of Plato was converted into a physical fact, and in accordance with the teaching of Plato in the *Laws*, the motion that animated the spheres was conceived of as their life or soul. The Prime Mover, or God, was an incorporeal spirit who animated the outermost sphere. How an incorporeal spirit could set a corporeal sphere in motion, a problem ignored by Plato in the *Laws*, proved a difficulty for Aristotle too. In the end he answers it by a metaphor. God moves the world in the same way in which a beloved moves the lover. It is the desire for God that sets the outermost sphere in motion. It was Aristotle who first discovered that " Oh, 'tis love, 'tis love, 'tis love that makes the world go round."

There is, however, the germ of a very great advance in method to be found in the *Physics*. Aristotle of course distinguishes physics from mathematics. The bodies studied by physics, he says, have in them geometrical entities, solids, planes, lines, and points. But the mathematician studies these geometrical entities in abstraction from things, while the physicist studies them as " limits of a physical body." This view is in accordance with the general opinion of thinkers in his day; but with Aristotle it is developed into a clear-cut distinction between matter and form which is fundamental for all subsequent thought. With this clear-cut distinction in mind Aristotle advances to a new conception of causation. The existence of anything depends, according to him, on four causes, the material cause, the formal cause, the efficient cause, and the final cause, that is to say, the stuff of which a thing is made, the pattern on which it is made, the thing which makes it, and the end or purpose for which it is made. It was characteristic of Aristotle to ponder carefully the results of his predecessors, and his four-fold doctrine of cause is a systematization of the ideas of earlier thinkers. The earth, air, fire, and water of the Ionians are examples of material causes; the Love and Hate of Empedocles are examples of efficient cause; the number-physics of the Pythagoreans had put the emphasis on the patterns of things, on the formal cause; and the demand of Socrates that it should be shown that whatever is is for the best, is satisfied by the final cause. But here the all-important distinction is to be observed, that Socrates had imagined the Forms to exist independently of things and to be the real reality, while for Aristotle the forms exist always as one aspect of a concrete thing.

It is this point especially that comes up again and again for discussion in the *Metaphysics*. The

Metaphysics is an enquiry into the nature of reality, and it is never assumed in this work that the supra-sensible world of forms, which was for Socrates the true reality, exists. The enquiry is directed rather to the question whether these forms exist at all, and if so in what sense. And the answer is that the forms do indeed exist but in inseparable association with matter. In other words, Aristotle begins his quest in the *Metaphysics* by asking whether the science of reality is concerned with perceptible substances or with supra-sensible forms. His conclusion is that it is concerned with both, for in reality the two are never disjoined. Form and matter are separable only in thought; what really exists is form incorporated in matter, " immattered form."

This is also the standpoint of the Aristotelian logic, a science created practically *ab initio* by Aristotle. Plato, it is true, had made a beginning in the *Sophist*. He had attempted to analyse what we mean by affirmative and negative propositions, and had suggested a few of the most universal notions under which we can classify all that exists. The most fundamental notion of all is *being*, common to everything that is. The next great heads of classification suggested by Plato are motion and rest, and likeness and difference. And with this as a start he had gone on to enquire how far these notions could or could not combine with one another. Thus being will combine with both rest and motion, for rest and motion both *are* ; but rest and motion will not combine with one another. But this analysis of Plato's had not gone beyond the discussion of the relation of logical categories to one another. The logic of Aristotle starts from the position laid down in the *Metaphysics*, that what really exists is the concrete thing. His logic is not a science of reality but a science of thought. Thought does not constitute things, it merely apprehends them. The distinction of matter and form is here also fundamental. The principle of individuation in things, what makes one chick-pea a separate entity from another chick-pea, is matter ; but what makes them both chick-peas is form, the form which is common to both. What the mind apprehends in things is the form, the intelligible aspect ; but this form, this intelligible aspect, has no existence apart from matter. The logic of Aristotle is the science of classifying the particular things of sense according to the forms inseparably associated with them.

This new conception of reality as immattered form inevitably required a fresh discussion of the old problem of reason and sensation as instruments of apprehending truth, and the lines of Aristotle's solution could not be in doubt. As in the external world he had united matter and form, so in the internal world he united sense and reason. With him there is no ultimate distinction between the two. The world that is the object of sensation is the same with the world that is the object of reason. The most elementary sensation, according to Aristotle, is not a purely passive thing, but has in it an active element, a power of discrimination, a certain degree of reason.

This synthesis of reason and sense was elaborately worked out in Aristotle's psychological treatise, *On the Soul*. Taking his cue from the Platonic teaching in the *Theætetus*, he developed the doctrine of a Common Sense, not in our meaning of the term, but in that of a single faculty of perception of which the special senses are parts. The separate senses each have special objects. Sight perceives colour, hearing sound, smell odours, taste sweetness and bitterness, touch (which Aristotle recognized as a compound sense) detects both hardness and softness, and heat and cold. But over and above these special objects of the individual senses there are what he calls " common sensibles," motion, shape, number, size and time, which are perceived by Common Sense rather than by any of the special senses. It is also the task of the Common Sense to compare the data of the special senses, to make conscious our sensory experiences so that we perceive that we perceive, and to exercise the faculties of imagination, memory, and expectation. To the Common Sense belong also the faculties of sleeping and dreaming, for the fact that the operation of all the senses together is suspended during sleep is proof of their being parts of one whole.

The elements of the psychological doctrine of Aristotle lie scattered throughout the pages of the later dialogues of Plato. But apart from the fact that Aristotle worked them up into a system and supported his views by an abundance of illustrative observations, there is the all important difference that Plato maintained a fundamental distinction between reason and sense and between their respective objects, while the purpose of Aristotle was to embed reason firmly in the senses as the active element by which we come to a knowledge of the forms that inhere in things. Inevitably, then, the Aristotelian doctrine of the soul differed fundamentally from the Platonic. The activity of the soul had been proof for Plato of its separateness from the body and its immortality. But Aristotle claims credit for being the first to recognize that mental phenomena are psycho-physical. He analyses very penetratingly the physiological basis of imagination, of memory, of dreaming, and of the various passions. This analysis undermined in him the Platonic belief in the immortality of the individual soul. For him soul and body are a unity, the soul being as it were the form of the body. The two can be separated only in thought, and the individual soul perishes along with the individual body. This statement, however, needs to be qualified to this extent that Aristotle recognized an element of thought that was purely logical and had no material content. Accordingly he distinguished the Mind, as the organ of pure thought, from the Soul which is the form of a living body, and taught that the Mind could enjoy an immortality of contemplation after the death of soul and body. This is a survival of Platonism which does not cohere logically with the rest of Aristotle's doctrine of the soul.

The effect of the new philosophical standpoint reached by Aristotle in his logical, metaphysical and psychological treatises was to clear the ground for the restoration of the experimental and observational study of nature. Socratic idealism by confining the field of true science to the forms of the supra-sensible world had been an obstacle to the quest for knowledge of nature, for a science of the phenomenal world. But, as has been well said, Aristotle's aim was to make the Idea capable of producing knowledge of appearances. And this he had achieved by his doctrine that the Idea or Form has no existence apart from the phenomenal world. The task of science was thus defined anew as the investigation of the material world in order to discover in it the universal forms. The tradition of Ionian science was resumed, but on a higher level. The Socratic and Platonic philosophy of mind had borne fruit in two new sciences, logic and psychology, the effect of which was to show how the mind by its capacity to apprehend forms could extract genuine knowledge from the flux of the phenomenal world.

In the last twelve years of his life Aristotle began to apply in various fields, but especially in biology, the method that he had hammered out in his middle period. That Aristotle had many predecessors in this field is certain ; but it seems certain also that he reconstituted and greatly extended the science. The principal works that have come down to us are the *History of Animals*, *On the Parts of Animals*, *On the Motion of Animals*, and *On the Generation of Animals*. In all in these works he mentions some 500 different species of animals ; and though of most of these his knowledge was second-hand and slight, yet he appears personally to have dissected some fifty types, a prodigious pioneering work. The novelty of these researches, at least in Athens, is implied in various passages of his writings in which the atmosphere of his lecture-room is conveyed to us with extraordinary freshness. As we read it becomes clear that an audience accustomed to identify science with *a priori* speculation about the divine stars has to be skilfully led, coaxed, and encouraged to take an interest in the details of biological research. One of these passages, a declaration by the head of the Lyceum of his faith in the new orientation he is giving to science, we shall quote at length. It is one of the golden pages in the literature of science.

" Natural objects fall into two great classes, the immortal ones that are without beginning and imperishable, and those that are subject to generation and decay. The former are worthy of honour and are divine, but are less within the reach of our observation, for all our speculations about them and our aspirations after knowledge of them can only in the rarest instances be confirmed by direct perception. But with regard to the plants and animals that perish, we are better off for coming to a knowledge of them, for we are inhabitants of the same earth. Anyone who is willing to take the necessary trouble can learn a great deal about all the species that exist. Both enquiries have their charm. Although in the case of the former we can achieve little owing to their being out of our reach, yet the veneration in which they are held imparts to knowledge of them a degree of pleasure greater than appertains to any of the things that are within our reach, as a lover would rather catch a random glimpse of his beloved than have a complete view of many other valuable things. But the latter, owing to our better and fuller acquaintance with them, have the advantage from the scientific point of view. Indeed their nearness to us and their kinship with us may be said to counterbalance the claims of divine philosophy. And as I have already expressed my views on the former subject, it remains for me to treat biology, omitting nothing so far as I can avoid it, however little or great be the honour in which it is held. For though there are aspects of the subject that are unpleasant to our senses, yet from the theoretical point of view the sight of nature at her constructive task affords incalculable satisfaction to those who are capable of coming to a knowledge of causes and who are philosophers by nature. For it would be irrational and absurd to take pleasure in images of natural objects, whether painted or modelled, because we detect also the skill of the artist at work, and to fail to love still more the constructions of nature herself, always supposing that we have the capacity to understand causes. Let us then not shrink like children from the investigation of the humbler creatures. In every natural object there is something to excite our admiration. You remember the story of Heraclitus, how when the strangers who wished to meet him halted in their approach on finding him warming himself at the kitchen fire, he bid them take heart and enter, saying that there also there were gods. So we too must take heart and approach the examination of every living thing without reluctance or disgust, for in everything is some part of nature, some element of beauty. Indeed it is in the works of nature most of all that we shall find purpose and freedom from indeterminacy ; and the end for which the thing has been made supplies the place of beauty in a work of art. Finally, if anybody should despise the study of other living creatures as unworthy of attention, let him think the same of himself, for it is not possible without great disgust to contemplate the elements of which mankind is made, the blood, the flesh, the bones, the veins, and all such parts." (*Parts of Animals*, I, 5, 1–7.) In this passage, precious alike for the biographer of Aristotle and for the historian of thought, the one point on which we can afford time to dwell is the teleological approach to science which is characteristic of Aristotle. For him the study of nature was the search for causes, and the most important cause was the final cause, the end or *telos* which nature has in view. In this way was Socrates' complaint that Anaxagoras had failed to develop the concept of mind in nature met and satisfied ; the one purpose of Aristotle was to trace the operation of mind in nature. But Aristotle did not teach (except in one passage, which perhaps should not be pressed) that there was a Mind outside the universe that consciously adapted means to ends. That was the naïve Socratic teleology. Aristotle's teleology is immanent. He was of opinion that the permanence of types in the animal world could not be explained merely by mechanical causation ; that the fact that from men sprang men, from frogs frogs, from bees bees, was proof that in the seed of these various stocks was some potentiality that could only be realized in one way, that necessarily imposed a certain form on matter ; that nature worked by a continual adaptation of matter to various forms, and that the manner of her working could not be understood without reference to the form which was to be produced. Applying the same line of thought to the various parts of animals, he sought in all his researches to reveal the adaptation of means to ends ; in other words, he habitually used function to explain structure. In this way he satisfied the Platonic demand that soul should be recognized as part of nature, though he denied the independent existence of soul. His biology is naturalistic but not mechanistic.

No point in Aristotle's biology has occasioned more discussion than his reliance on the concept of final cause. A few biologists even at the present day still find this an indispensable concept, though most have abandoned it and all look with suspicion on the tendency it encourages to give facile pseudo-explanations of the facts which one is seeking to understand, a tendency exhibited by Aristotle himself in his physical treatises. To say that God put our mouth just under our nose so that we might enjoy the smell of our food, as Socrates did, or that giraffes have got long necks because they badly wanted to eat the tops of trees, as has been suggested in modern times, is teleology of a sort and is certainly an offence to serious scientists. But this was not the teleology of Aristotle. With him teleology was not a substitute for observation, but a guide to research and an ideal to be reached. Aristotle above all was not an arm-chair biologist. It may be allowed to quote proof of this. In his treatise *On Generation* he writes thus about bees : " The facts have not yet been sufficiently grasped ; if they ever are, then credit must be given to observation rather than to theories, and to theories only in so far as they are confirmed by the observed facts."

The great general result of Aristotle's biological researches was the establishment of a *scala naturæ*, a method of classifying all living creatures, not superseded until the time of Linnæus. It is reproduced here from W. D. Ross's *Aristotle* :

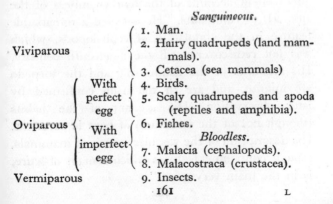

Sanguineous.

Viviparous
1. Man.
2. Hairy quadrupeds (land mammals).
3. Cetacea (sea mammals)

Oviparous — With perfect egg
4. Birds.
5. Scaly quadrupeds and apoda (reptiles and amphibia).

With imperfect egg
6. Fishes.

Bloodless.
7. Malacia (cephalopods).
8. Malacostraca (crustacea).

Vermiparous
9. Insects.

L

Bloodless.

| Produced by generative slime, budding or spontaneous generation. | 10. Ostracoderma (molluscs other than cephalopods). |
| Produced by spontaneous generation. | 11. Zoophytes. |

Certain particular results (also quoted here from Ross's *Aristotle*) have been much admired. " He recognized the mammalian character of the cetaceans—a fact which was overlooked by all other writers till the sixteenth century. He distinguished the cartilaginous from the bony fishes, and described them with marvellous accuracy. He describes carefully the development of the embryo chicken, and detected on the fourth day after the laying of the egg the presence of the heart ' like a speck of blood in the white of the egg, beating and moving as though endowed with life.' He gives an excellent account of the four chambers of the stomach of ruminants. He detected a remarkable feature in the copulation of cephalopods, which was not rediscovered till the nineteenth century. His accounts of the fishing-frog and the torpedo are minute, and are in the main confirmed by later observation. His account of the habits (though not of the structure) of bees is excellent. His description of the vascular system of mammals, though containing features which remain obscure, is in the main very good."

Critics have, of course, been able to point to many mistakes in Aristotle's work, the most notorious of which is, perhaps, his rejection of the teaching of Alcmæon that the brain is the seat of sensation, and his choice of the heart for the performance of that function. This was, of course, a retrograde step, but nobody who considers the reasons advanced by Aristotle for his choice will consider the mistake of much importance in comparison with the evidence they provide of his indefatigable industry and the general excellence of his method. It is to his credit that he was not content to accept without examination the results achieved by an early researcher some 150 years before his time.

His reasons for rejecting Alcmæon's view were as follows :

(1). The brain is insensible to external mechanical stimulation. If the brain of a living animal be exposed it may be cut without the animal exhibiting any sign of distress. In Aristotle's day the function of the nerves was not understood, so that this experiment seemed to tell against the theory of Alcmæon.

(2). In invertebrate animals the cephalic ganglia which take the place of the brain are generally too small to be observed with ease by the naked eye. They eluded the observation of Aristotle, who thought that these animals had no brain ; but he knew that they were sentient, and therefore could not connect sentience with the brain.

(3). Like Hippocrates he wrongly regarded the brain as bloodless, and much experiment had convinced him that only sanguineous parts are sensitive.

(4). He thought it manifest to observation that there is no anatomical connection between the brain and the sense-organs.

(5). He thought he could discover connecting links between the sense-organs and the heart.

(6). The heart is the centre of the vascular system and of the vital heat (a Hippocratic theory in which Aristotle believed).

(7). The heart is the first part to become active in the embryo and the last to stop working in the dying animal. It is therefore likely to be the seat of sensibility, the most essential characteristic of animal life.

(8). The action of the heart is augmented or diminished when intense pleasure or pain is experienced.

(9). Loss of blood causes insensibility.

(10). The heart is the central organ in the body, which seems to fit it to be the organ of the central sense.

Finally a word or two must be said about Aristotle's researches other than biological. When he was still a student at the Academy his passion for reading had already been the occasion for comment.

At the Lyceum he collected the first great library in Europe, and this subsequently became the model for those at Alexandria and Pergamum. This collection of books was not a mere luxury with Aristotle ; it was the essential pre-requisite of one of the most original features of the activity of the Lyceum. Now for the first time the historical approach to problems of philosophy and science began to be applied. In his *Metaphysics* Aristotle begins with a rapid sketch of the previous history of thought designed to present what is original in his own views in its true perspective. The activities of other members of the school were directed to the same end. A series of historical studies in philosophy and science were undertaken which, though they have not themselves survived, lie at the basis of most of our knowledge of the thought of ancient Greece. The task of writing the history of physics and metaphysics was entrusted to Theophrastus, his favourite pupil and the future head of the Lyceum, and was executed in eighteen books. To one Eudemus was assigned the province of mathematics, that is to say arithmetic, geometry, astronomy, and probably theology, the connection of which with mathematics has already been commented upon. The historian of medicine was one Menon. Theophrastus also undertook and carried through the composition of an extensive work on botany, which did for plant life what Aristotle in his various biological treatises had done for animals.

The great effort at systematizing and organizing knowledge was given an extension beyond the range of natural science and made to embrace the social sciences and the history of culture. The periodic festivals and public games were an important feature of Greek life. Aristotle compiled a list of victors at the Pythian games and made a collection of the records of the dramatic contests at the two Athenian festivals where they took place. Plato, under the belief that drama appeals to the lower side of man's nature, both in his *Republic* and his *Laws* banished the tragedians from his model states. Aristotle's humble service to the art which Plato condemned laid the foundation of our knowledge of the development of Athenian tragedy and comedy; and the philosophic justification for this attention to drama was elaborated in a brilliant treatise on poetry in which, without mention of his revered master's name, all the Platonic attacks on tragedy are countered and destroyed. A like tolerance was extended to Homer who, on account of his secular temper and irreverent handling of the Olympian gods, was another of Plato's *bêtes noires*. Both in his treatise on poetry and in his *Homeric Problems* Aristotle gave the great poet the place that was his due, and also constituted for the first time the science of philology which passed to the Museum at Alexandria along with the rest of the Aristotelian tradition. In politics also his contribution was of importance, but more perhaps owing to his historical method than to any exceptional power of penetrating into the nature of human society or seeing further than the narrow bounds of the Greek city state. With the collaboration of members of his school he described the constitutions of 158 city states. Of this vast pioneering work in political research little has survived, although by great good fortune the first of the collection, the constitution of Athens described by Aristotle himself, has in modern times come to light.

CHAPTER VIII

THE ALEXANDRIAN AGE

AFTER the death of Aristotle the renown of Athens as a centre of scientific research was rapidly eclipsed by Alexandria. Here Ptolemy, one of the generals of Alexander the Great, had established himself as head of portion of the vast empire Alexander had won ; and the dynasty he founded in the new capital of Egypt, where a Greek court ruled over the ancient peoples of Egypt, and where a large Jewish element contributed to the cosmopolitan character of the population, patronized learning with lavish generosity. The Museum which the Ptolemies founded and maintained in Alexandria rapidly became the centre of a scientific movement that might have transformed antiquity into a semblance of the modern world. Ancient society halted on the threshold of a machine age. But before we direct our attention to the Egyptian capital we must give a last glance at Athens.

The pre-eminence of Athens in the fourth century as a centre for higher education rested chiefly on three great schools, that of the rhetorician Isocrates, the Academy of Plato, and the Lyceum of Aristotle.

The school of Isocrates, the first to be opened, dates probably from 393 B.C. It was humanistic rather than scientific in character. The founder was not interested in metaphysical questions and was sceptical of the value of higher mathematics ; but his extraordinary gift of eloquence and his talents as a publicist attracted to his school those whose wealth and birth marked them out for careers as rulers or high officials in the various Greek states. His public discourses on topics of cultural and national importance which are still extant are proof of the high level of general culture he was capable of imparting.

In contradistinction to the school of Isocrates the Academy, founded about 387 B.C., was philosophic and scientific in character. A preliminary training in mathematics was required for entrance, and under the direction of Plato Pythagorean mathematics were reconstituted on a new basis, principally through the work of two men, Theætetus and Eudoxus, the ripe fruit of whose endeavours was garnered by Euclid at Alexandria. Within the bosom of the Academy were born also the new tendencies that were represented by Aristotle and the Lyceum. Nor must it be forgotten that the Academy had an ultimate political goal, the reconstitution of society by the reconstitution of the individual. The activities that found expression in the lifetime of the founder by the publication of the *Republic* and the *Laws* were recognized after his death by the appeal several times addressed to the Academy by one or other city state for help in drafting its constitution.

The career of the Lyceum, in contrast to that of the Academy, was meteoric in its brilliance and brevity. Its activity was practically confined to the lifetime of its founder and that of his immediate successor, Theophrastus ; the Lyceum was then, in effect, reborn in the Museum at Alexandria. But the Academy continued to function at Athens, producing a long line of mathematicians, of whom some of the more famous are Speusippus, Xeno-crates, Dinostratus, and Theudius, and developing a sceptical, critical outlook which long exercised its sobering effect on the thought of the ancient world. In all the Academy had an uninterrupted existence of over 900 years, being eventually closed by the Emperor Justinian in the third decade of the sixth century A.D. ; and during the whole of the Alexandrian Age its existence together with that of the rival Stoic and Epicurean schools enabled Athens to retain in the philosophic world the pre-eminence which Alexandria had captured in the scientific field.

Before the palm passed to Alexandria, however, two remarkable scientific achievements were made on Attic soil. Heraclides of Pontus (388–310 B.C.) settled in Athens some time in the second half of the fourth century and there propounded for the first time two revolutionary discoveries in astronomy that were destined to immortalize his name. It will be remembered how Plato had set it as a problem to the members of the Academy to determine " the uniform and ordered movements by the assumption of which the apparent movements of the planets could be accounted for " ; and how Eudoxus had provided the geometrical solution of the problem, which Aristotle had afterwards transformed into a mechanical model of the structure of the universe. Heraclides now produced a much truer solution of Plato's problem. He discovered that the planets Venus and Mercury (whose orbits lie closer to the sun than that of the earth, and which consequently are never seen at any great distance from the sun) revolve, not about the earth, but about the sun. And of the two possible alternatives that could account for the appearance of the daily revolution of the starry heaven, namely that it actually revolves or that the appearance of rotation is produced by the revolution of the earth about its own axis, he decided, correctly, in favour of the latter. This was the true line of development, which ultimately not only saved the appearances but revealed the facts.

The second achievement was in medicine. Diocles, originally of Carystus in Euboea, made his headquarters in Athens, and although his works are lost, he is known to have made remarkable advances, especially in human embryology. He claimed to have examined the human fœtus at various stages of its growth, and at 27 days to have found traces of the head and spinal column, and at 40 days to have been able to distinguish the form as human.

But with the rise of Alexandria the scientific pre-eminence of Athens was gone. Typical of the change is the career of Strato of Lampsacus, who succeeded Theophrastus in the headship of the Lyceum. Too little is preserved of the researches of this remarkable man, but he is known to have taken up again the physical rather than the biological side of natural philosophy and to have attempted to rebuild the physical theories of Aristotle on the basis of the atomism of Democritus. Unhappily for the position of Athens in the history of science and for the continued existence of the Lyceum, Strato did not end his career where it had begun. He was summoned to Alexandria by the founder of the Ptolemaic dynasty. With the departure of the head of the Lyceum the whole scientific activity of that institution was transferred to the Egyptian capital.

Strato's revival of atomism is a step of such importance that although our materials are scanty it is proper to throw as much light upon it as we can. Plato's hostility to Democritus had been absolute; he never mentions his name, and is reputed to have wished to burn all his books. But Aristotle, who never falls a victim to intolerance, had insight enough to appreciate his greatness and alludes to him in a significant passage in the *Parts of Animals*. In the passage in question Aristotle discusses the approximations the older philosophers made to the ideas of formal and final cause. He observes how men like Empedocles were sometimes compelled to give an unconscious adherence to the idea of form. For example, when Empedocles attempts to give an account of organic substances such as bone and blood, the four elements are sufficient to give him only the material cause, and to complete his explanation he lays emphasis also on the *proportion* of the mixture, thus unconsciously falling back on what Aristotle calls formal cause.

The same happened with the idea of final cause. The ancients had not succeeded in realizing it clearly, but were obviously in need of it. Thus the physicists had attempted to account for the hollow cavity of the stomach by saying that it was caused by the rush of the element of water inside the earthy body, and for the nostrils by saying that they were caused by a violent outbreak of enclosed air. This sort of mechanical explanation Aristotle thought insufficient ; his teleology would complete it by the idea that the perfect form of man, including the possession of digestive and respiratory apparatus, is part of nature's eternal plan which she continually imposes on matter. It is in this connection that his famous reference to Democritus occurs. " Democritus was the first to seize the importance of final cause, but without realizing its importance for physical enquiry ; with Socrates the idea of final cause was indeed developed, but the enquiry into nature came to an end and philosophers turned aside to ethics and politics."

Three points in this quotation are worthy of note. In the first place we have an openly expressed regret for the Socratic interruption of natural philosophy ; secondly, the somewhat surprising statement that the great determinist Democritus had realized the importance of final cause ; thirdly, the criticism that he had not grasped its importance for physical enquiry. The last seems to be precisely the point on which Strato reverted from Aristotle's position to that of Democritus. The idea of final cause retained its place in biological research, but from the time of Strato on it is banished from physics. The Socratic moratorium on physical enquiry had come to an end.

In Alexandria, whither Strato had been summoned by Ptolemy, the organization of studies was on a hitherto unprecedented scale, though the actual foundation of the Museum may not have taken place till the reign of the second Ptolemy. The Museum, or Temple of the Muses, was nominally under the charge of a high-priest—a concession to the long-

established tradition of priestly culture in Egypt—
but it was in essence a research university which
also did some teaching. Now for the first time,
in what was culturally a European city, learning
and research were adequately supported by the
state. It is true that the maintenance of royal
libraries was far from being without precedent.
The tyrants Peisistratus of Athens and Polycrates
of Samos had had libraries in the sixth century in
imitation of the practice of the east, where libraries
had long existed—in Babylonia, Assyria, Phœnicia,
Jerusalem, and Cappadocia. But not only did the
Museum possess the most important library that
ever yet had been or could have been gathered
together (the nucleus of it was formed by the
library of Aristotle himself, it soon numbered
among its treasures such priceless possessions as the
official text of the Athenian tragedies, and the total
of volumes rapidly mounted above the half-million),
it also contained many other aids to research.
There were rooms for lectures and study, dissecting
rooms, an observatory, a zoo, a botanical garden—
in a word all the material requirements for the
anatomical, astronomical, biological, botanical and
philological studies that were destined to make such
rapid progress there. Nor were the men lacking.
With abundant financial resources the Ptolemies
had no difficulty in attracting to their institution
the best brains. The king paid the staff, and

among the hundred or so regius professors that the establishment supported a truly remarkable number have left their names to posterity as benefactors of mankind. The institution lasted in all some 600 years, but underwent various vicissitudes, the worst disaster it experienced being the burning of the library during Julius Cæsar's Alexandrine campaign. But one hundred years earlier, in 145 B.C., the Museum had been temporarily broken up, and the period of its greatest glory brought to an end, by the violence of Ptolemy Euergetes II, who on returning victorious from a civil war gave Alexandria up to sack. The vital period of its existence is the first 150 years. Within that period fall the works of the three great classical mathematicians, Euclid, Archimedes, and Apollonius. Within that period a succession of great astronomers and geographers had taught mankind how to map the heavens and to map the earth, the essential preliminary to the organization of civilization on a large scale. Among the biological sciences, anatomy and physiology had made sensational progress. Mechanics had indefinitely multiplied the forces at the disposal of man. And grammarians had analysed the structure of language and established sound principles of textual criticism, without which the record of civilization cannot be preserved and progress becomes uncertain and blind. The means to a greatly increased control of his destiny lay ready to man's hand had he been capable of using them. Unhappily, the harvest was poorly reaped.

Since the advances in medical science were all achieved within the period 300–250 B.C., it will be convenient first to deal with them. They were the work of two men, both Asiatic Greeks, who by extensive operations on the bodies of men and animals brought anatomy and physiology to a pitch that was not seriously improved upon until the sixteenth century of our era. The anatomical discoveries are principally associated with the name of Herophilus of Chalcedon. His works are lost, but from the tradition that survives it appears that his best work was done on the brain. The *torcular Herophili* is called after him ; he described also the meninges, distinguished the cerebrum from the cerebellum, and identified a depression in the fourth ventricle, the *calamus scriptorius*, which is the Latin equivalent of the name he gave to it. He was also the first to grasp the nature of nerves other than those of the special senses, and to divide the nerves into motor and sensory. The result of his discoveries was that the Aristotelian confusion of the functions of the brain and heart was quickly corrected, and opinion reverted to the view of Alcmæon, that the brain is the central organ of the nervous system and the seat of the intelligence.

The physiological system of Erasistratus of Chios is superficially less impressive, for it depends on the concepts of the vital and animal spirits, which, since Harvey's demonstration of the course of the circulation of the blood, have definitely fallen back into the limbo of dead things. But the system was based on extensive investigation into the vascular system of men and animals. Among the vessels described by Erasistratus Dr. Singer lists the Aorta, Aorta descendens, Pulmonary artery, Intercostal arteries, Hepatic arteries, Renal arteries, Gastric arteries, Pulmonary veins, Vena cava, Azygos vein, and Hepatic veins. Erasistratus, of course, distinguished the veins from the arteries, and even his mistaken view, that the arteries contained air not blood, involved knowledge of the capillary connections between the venous and arterial systems.

Although the works of the Alexandrian doctors have perished, it must not be forgotten that it is largely to the multiplication and distribution of books at Alexandria that we owe the preservation of such of the learning and literature of Greek antiquity as has survived, the Hippocratic collection included. Here at Alexandria were collected, criticized, corrected and edited the masterpieces of Greek literature and science, prose and verse. Hand in hand with this critical work, and growing naturally out of it, went the formulation of the science of grammar along lines that are still serviceable to-day. The first extant Greek grammar, which of course represents the culmination of a long tradition, is that of Dionysius Thrax or Denys of Thrace, who flourished about 120 B.C. and was a pupil of Aristarchus, librarian at Alexandria and a famous textual critic. Denys's grammar is only sixteen pages long in its modern printed form, but it is admirably orderly and comprehensive. Beginning with a definition of grammar as " the practical knowledge of the usage of writers of poetry and prose," it then divides the subject into six parts which contain a good deal that we do not now include under grammar. The six parts are : accurate reading, explanation of figures of speech, exposition of rare words and subject-matter, etymology, doctrine of the regular grammatical forms, and, lastly, criticism of poetry " which is the noblest part of all." Then, in accordance with the above division of the subject, the author proceeds to deal with accentuation, punctuation, letters, syllables, parts of speech, declension, and conjugation. There is here so much sound doctrine clearly stated that it is not surprising that for some thirteen centuries it remained the standard text-book on its subject.

Celebrated, however, as this Alexandrian text-book of grammar is, its fame is eclipsed by another product of the same place, the *Elements* of Euclid. This, the most famous text-book in the history of the world, was written by a man of whom we know little more than his name and the fact that he flourished during the lifetime of the first Ptolemy.

His book, which resumed and completed the centuries-old tradition of Greek geometry and arithmetic (i.e. theory of numbers) at once became a classic. Archimedes, who was born about 287 B.C., already cites it by book and proposition ; and it was still the current text-book in elementary geometry in English-speaking countries at the beginning of this century. In addition to the thirteen books of the *Elements* other works of Euclid are still extant : the *Phænomena*, a work on spherical astronomy, and the *Optics*, a treatise on perspective. The latter subject seems to have been forced into prominence by the requirements of the stage.

Astronomical work at Alexandria was not merely theoretical but observational, and it soon led to an advance that anticipated the Copernican revolution. The author of this advance was Aristarchus of Samos, a pupil of Strato of Lampsacus, and like him a researcher in physics. The work in which he announced his revolutionary astronomical hypothesis is lost, but we are fortunate enough to possess a summary account of his views in the writings of his younger contemporary Archimedes. " His hypothesis," Archimedes tells us, " is that the fixed stars and the sun remain unmoved, and that the earth revolves about the sun in the circumference of a circle, the sun lying in the middle of its orbit." It is much to be regretted that we do not possess the arguments by which Aristarchus supported his view.

Aristarchus also greatly extended current ideas about the size of the universe, for Archimedes goes on to report that according to him " the sphere of the fixed stars, situated around the same centre as the sun, is so great that the circle in which the earth revolves bears a proportion to the distance of the fixed stars the same as the centre of a sphere bears to its surface." This is equivalent to the statement that the distance of the fixed stars is so great that the diameter of the earth's annual orbit about the sun is negligible in comparison. Aristarchus's reason for making such an extravagant statement is that astronomical instruments had not yet reached a sufficient degree of precision to enable him to detect the phenomenon of parallax.

With the Alexandrian astronomers the goal to be attained was, of course, not the Platonic one of solving problems in spherical geometry, but that of finding out the physical facts. The one writing of Aristarchus that has come down to us is entitled *On the Sizes and Distances of the Sun and Moon.* It was written before he had arrived at his heliocentric view of the universe, but as an extant example of the astronomical science of the day it is worthy of further discussion even within the narrow limits of our book.

The methodical arrangement of the work is

worthy of the age that had just produced the *Elements*. Aristarchus begins by stating the six hypotheses or assumptions on which his argument is based and on the truth or falsehood of which it stands or falls. They are :

1. That the moon receives its light from the sun.

2. That the earth is in the relation of a point and centre to the sphere in which the moon moves ; i.e. he proposes to neglect the diameter of the earth as negligible in comparison with the diameter of the moon's orbit.

3. That, when the moon appears to us to be halved, the great circle which divides the dark and bright portions of the moon is in the direction of our eye. This, as a diagram would at once show, means that the centres of the sun, earth, and moon form a right-angled triangle with the right angle at the centre of the moon.

4. That when the moon appears to us to be halved its angular distance from the sun is 87 degrees. This is slightly out. The true angle is over 89 degrees.

5. That the breadth of the earth's shadow is that of two moons.

6. That the moon subtends an angle of 2 degrees. This is again wrong. Archimedes reports a later estimate of Aristarchus, namely $\frac{1}{2}$ a degree, which is correct.

" We are now in a position," Aristarchus proceeds, " to prove the following propositions :

1. The distance of the sun from the earth is greater than eighteen times but less than twenty times the distance of the moon from the earth.

2. The diameter of the sun has the same ratio as aforesaid to the diameter of the moon.

3. The diameter of the sun has to the diameter of the earth a ratio greater than 19 : 3 but less than 43 : 6."

Eighteen propositions are then proved, of which numbers 7, 9, and 15 cover the three mentioned above as being those which it is the special object of the treatise to establish.

The method of this treatise is unexceptionable, namely, mathematical theory applied to the data of observation. That the data are inaccurate is not so important. The Alexandrians were well aware of these imperfections ; repeatedly in his extant writings Archimedes observes that neither eyes, hands, nor instruments sufficed to give him the accuracy he required, and in one of his treatises he devotes five or six pages to the description of an ingenious device for narrowing the margin of error in determining the angle subtended at the eye by the sun's disc. Aristarchus's estimate (hypothesis 5) of the ratio of the earth's shadow to the diameter of the moon was made by timing the passage of the moon through the shadow of the earth at an eclipse. His estimate, probably made with a water-clock, is too small.

One surprising deficiency in observation raises a question of such interest that it demands a short digression. Proposition 8 of the treatise *On the Sizes and Distances of the Sun and Moon* is as follows : " When the sun is totally eclipsed the sun and moon are then comprehended by one and the same cone which has its vertex at our eye." Aristarchus uses this observation as the justification for his view that the diameters of the sun and moon subtend the same angle at our eye. But evidence was already available tending to prove that the distances of the sun and moon from the earth are not constant. A late Greek author, Simplicius, writing in A.D. 536, collects evidence for variations in the distances of the heavenly bodies from the earth, and then proceeds thus : " There is also evidence for the truth of what I have stated in the observed facts with regard to total eclipses of the sun ; for when the centre of the sun, the centre of the moon, and our eye happen to be in one straight line, what is seen is not always alike ; but at one time the cone which comprehends the moon and has its vertex at our eye comprehends the sun itself at the same time, and the sun even remains invisible to us for a certain time, while again at another time this is so far from being the case that a rim of a certain breadth on the outside edge is left visible all round it at the middle of the duration of the eclipse. Hence we must conclude that the apparent differences in the sizes of the two bodies observed under the same atmospheric conditions is due to the variation in their distances."

Simplicius thus proves the insufficiency of the observational data of Aristarchus to support the conclusion he bases on them. He also adds the interesting information that variations in the sizes of the planets, indicating variations in their distances from us at different times, were known as early as Aristotle and before him. It looks as if we had here an example of the danger of neglecting evidence that does not fit in with a preconceived theory. If Aristotle had allowed due weight to these observations, it would have precluded his invention of his system of concentric spheres based on the doctrine of the unchangeability of the heavens and on reverence for the sphere as the perfect figure, which would have been a real gain for science. Nor would the valuable argument of Aristarchus have suffered by the admission that at Solar eclipses the moon's disc sometimes overlaps the sun and sometimes fails to cover it. In fairness to Aristotle, however, it should be stated that his *Metaphysics* contains an explicit caution that his astronomical views are provisional and subject always to more accurate observations by specialists.

Aristarchus had attempted only to determine within certain limits the ratios of the diameters of the sun and moon to the diameter of the earth ; an accurate estimate of the diameter of the earth in standard units of measurement was still lacking. Aristotle, for instance, quotes with approval a figure that is almost twice the correct length. The desideratum was supplied by Eratosthenes, a younger contemporary of Aristarchus. He was librarian at Alexandria and the greatest geographer of antiquity. His calculation rested upon a combination of astronomical observation with direct measurement of a portion of the earth's surface. At the summer solstice at Syene the sun at noon, as can be inferred from the fact that it illuminates the bottom of a deep well, is vertically overhead. At Alexandria at the same time the distance of the sun from the zenith is $\frac{1}{50}$ of the circumference of the heavens. These two towns were judged by Eratosthenes to be situated on the same meridian and to be 5,000 stades apart, both of which data are approximately correct. 5,000 stades is therefore $\frac{1}{50}$ of the circumference of the earth. Worked out in our units this gives a length for the circumference of about 24,662 miles, a very close approximation ; it implies a value for the diameter which is only 50 miles less than the true polar diameter.

As a strictly scientific geographer Eratosthenes required that astronomical data should be employed for plotting the map of the world according to parallels of latitude and meridians of longitude, and as a true heir of the Aristotelian tradition he prefaced his own contribution to geographical knowledge with a history of the subject from Homer downwards to his own date.

The Egyptians had not done very much for geographical science. As early as 3200 B.C. we hear of a maritime venture under King Snefru, when an expedition went with 40 ships of 100 cubits length to fetch cedar-wood from Byblus in Phœnicia. But the exploration of the Levant progressed no further with the Egyptians ; and the most famous expedition undertaken under Egyptian auspices was entrusted to foreigners, the Phœnicians, who, according to a not impossible report, circumnavigated Africa in 600 B.C. The first extensive exploration of the Levant was the work of the Minoans, who had their headquarters in the island of Crete. They had probably a good knowledge of the eastern Mediterranean before 1500 B.C., and had extended their voyages to Sicily and South Italy by 1200 B.C. It was the Phœnicians, however, in the first millennium B.C. who first opened up the Mediterranean from end to end ; while about 500 B.C. the Carthaginian Hanno coasted down the west coast of Africa as far as Sierra Leone, to within 8 degrees of the equator. Close on the heels of the Phœnicians followed their rivals the Greeks. The effective discovery of the Black Sea was the work of the Greeks of Miletus. Exploration was begun about 800 B.C. and by 650 B.C. there was a heavy fringe of Greek colonies all round the Black Sea coast. It was for this city of explorers and merchants that Anaximander constructed his map. It was not from Miletus, however, but from the neighbouring town of Phocæa, that the great enterprise was organized that eventually turned the Mediterranean into something like a Greek lake. Probably as early as 850 B.C. colonists from Phocæa occupied the Italian site of Cumæ near Naples, and by the end of the seventh century numerous Greek settlements, mostly Phocæan in origin, dotted the western Mediterranean. The most westerly of these settlements was at Mænace near Malaga in Spain, the most famous was at the first site of civilization in France, Marseilles.

It was from Marseilles that a Phocæan sea-captain, Pytheas, about 300 B.C. made one of the great voyages of antiquity. Eluding the vigilance of the Carthaginians, who still at that date endeavoured to keep the Atlantic as their preserve, he slipped through the Pillars of Hercules and coasted north. His chief objective was the tin mines of Cornwall, which he visited and well describes. Pytheas was an educated man and a skilled astronomer, and his voyage was rich in scientific results. He was capable of discovering that the Pole star is not situated exactly at the pole, and of determining the latitude of Marseilles to within a few minutes of the correct figure. His accurate observations gave later geographers their reference points in plotting the map of northern and central Europe ; and it was due to information derived from him that Eratosthenes succeeded in locating Ireland in its right position relative to Britain.

Scientific voyagers like Pytheas were capable of calculating latitudes astronomically. Longitude, however, remained a matter of dead-reckoning. The compass and the magnetic pole were unknown and the captain had to rely simply on a guess made from the average speed of his craft with allowance for special conditions. The lengths of the Black Sea and of the Mediterranean were always greatly over-estimated in antiquity.

Conditions were similar on land. There the sun-dial and the polos enabled the scientist to determine latitude with fair accuracy ; but distances as a rule were measured directly by " steppers " such as accompanied the army of Alexander in its march to the east. Such were the instruments and such the data on which the ancient geographers relied for the construction of their maps. With regard to the first Greek map, that of Anaximander, details of its construction are lacking ; but a successor of his, Hecatæus, likewise of Miletus, who made a map about 510 B.C. to accompany a description of the world, grouped the lands he described in a symmetrical arrangement round the Mediterranean Sea. Maps constructed on this plan became common in the fifth century, and Aristotle in the fourth century is known to have had a collection of maps at the Lyceum. Finally Eratosthenes, gathering together the results of this long effort and completing it by his superior science, estimated the habitable region of the earth as an area of about 9,000 miles in length and 4,500 miles in breadth, which he divided into rectangles by lines parallel to the equator and by meridians. His two main axes intersected at the island of Rhodes, and the extreme meridians lay at the Pillars of Hercules in the west and the Ganges in the east. Thus the principles of scientific map-making were determined with a serviceable degree of accuracy. But it should be added that at no time in antiquity, owing to the lack of a sufficiently organized and extensive effort, did the detail of the habitable world receive the precise determination which their science permitted. Distances continued to be measured by dead-reckoning and were generally inaccurate.

As the map-makers of Miletus in the sixth and fifth centuries were certainly not uninfluenced by the practical needs of their trading and colonizing community, so there is no room for doubt that the Ptolemies who financed the great scientific effort of Alexandria were interested in the practical application of its result. Nevertheless, a feeling was prevalent among the Greeks that science ought to be useless, and the purely theoretical side of their activities is generally most strongly stressed in the ancient writers. This is notably illustrated in the career of Archimedes, by general acclaim the greatest mathematician and mechanist of antiquity, and among the greatest scientists of all time. Although a native of Syracuse he belongs to the Alexandrian school, for he had studied in Alexandria and always kept in close touch with scientific circles there. But his life was spent mainly in Sicily, where as kinsman and friend of King Hiero, he preferred to adorn the city of his birth and assert his local patriotism by the use of the Doric dialect even for his scientific masterpieces. Incidentally it may be remarked that the style of Archimedes is worthy of the man ; it is a model of simplicity and precision.

That Archimedes had practical mechanical genius of the first water is proved by the many inventions that stand to his credit. In his youth he constructed a planetarium, which in the doubtless too sweeping phrase of Cicero " with a single motor reproduced all the unequal and different movements of the heavenly bodies." In Egypt he was credited with the invention of a spiral pump for raising water from the Nile. He is reputed to have detected adulteration of the gold in Hiero's crown by a method involving the discovery of the principle of specific gravity. He launched a great galley for Hiero by a system of compound pulleys. Above all, he defended Syracuse against Roman attack by an array of military engines at that time without precedent. Nevertheless it was on his pure theory that he prided himself most. On his tomb he wished to be represented a sphere inscribed in a cylinder together with the formula he had discovered for the volume by which a cylinder exceeds its inscribed sphere.

Interesting evidence on the attitude of the Greeks to mechanics is given by Plutarch in his account of Archimedes in his *Life of Marcellus*, the Roman general who captured Syracuse. The celebrated and popular art of mechanics, Plutarch tells us, was first originated by Eudoxus and Archytas, contemporaries and friends of Plato. These men, faced by theoretical problems that defied their powers of analysis by argument and diagram, fell back upon ingenious mechanical constructions that solved their problems in a practical way. Plato was distressed at this and accused them of " destroying all the good of geometry," which, under their influence, had escaped like a runaway slave from the incorporeal and intelligible to the sensible, and had resorted to the use of material bodies requiring vulgar practical manipulation. In this way, Plutarch continues, mechanics was hissed off the stage ; it was completely divorced from geometry, suffered persistent neglect from the philosophers, and sank to the level of a military art.

One might suppose from the terms he uses that Plutarch, writing about A.D. 100, had risen superior to the prejudices of a former age and regretted the expulsion of mechanics from the company of the liberal arts. But this does not appear to have been the case. He was unhappily too loyal a Platonist for that. For he goes on to praise Archimedes for his lofty contempt for his practical achievements. Although his engineering feats had won him a reputation for superhuman ingenuity, yet, Plutarch tells us, so lofty was his spirit, so pregnant his genius, so rich his store of theory, that he refused to leave behind him any treatise on mechanics or on any art whatsoever that touched on the practical. Such interests he regarded as ignoble and vulgar and preferred to concentrate all his powers on subjects whose charm and excellence were uncontaminated by application to human needs.

This passion for pure science, and the failure, so often commented upon, of their theoretical progress to give rise to a more mechanized civilization, seem to be due in some degree to the structure of Greek society. Their slaves were their machines, and so long as they were cheap there was no need to try to supersede them. The supply of slave labour seems to have outlasted the hey-day of ancient science, thus robbing it of the economic motive for the invention of machines. It would be a nice question to determine whether a humanitarian motive has ever operated to produce mechanical inventions. It was certainly unlikely to do so in antiquity. The voice of pity was, indeed, not wholly inaudible ; but a convenient belief prevailed that nature had cast men in two moulds, the thinker and the manual worker. On this division society was based. The producers were regarded by some as unworthy of citizenship. This is the view of Aristotle, who would limit citizenship to the ruling class and the soldiers. For his exclusion of the producers from the citizen body he produces a striking argument. The producers, he says, are necessary to the state but they do not form part of it, just as a field is necessary to maintain a cow but is not part of the cow. In a society where such arguments carried weight science was the preserve and privilege of the economically independent and was not felt to have a social function, but to be of value primarily as a discipline for the individual soul of one designed by nature to be a thinker.

We need feel no surprise, then, that it was on the theoretical side of his achievement that Archimedes prided himself. His contribution both to theoretical mathematics and theoretical mechanics was enormous. Among his extant works are the following : *On the Sphere and the Cylinder* (two books), *Measurement of a Circle*, *On Conoids and Spheroids*, *On Spirals*, *Quadrature of the Parabola*, *The Sand-reckoner*, *On Floating Bodies* (two books), *On Plane*

Equilibriums (two books). *The Sand-reckoner* is a popular treatise addressed to Gelo, the successor of Hiero, showing how the Greek numerical notation may be adapted to the expression of any number no matter how great. In the two books *On Floating Bodies* the science of hydrostatics is constituted. It is in this treatise that the famous proposition is announced " that a body immersed in a liquid loses weight equal to the weight of the liquid displaced." The same treatise also states and proves the proposition that " the surface of any fluid at rest is a sphere the centre of which is the same as that of the earth." The treatise *On Plane Equilibriums*, basing itself on a few postulates such as that " equal weights at equal distances balance," advances with magisterial logic to the fundamental theorem that " two weights balance at distances reciprocally proportional to their magnitudes." It was in the confidence of the strength of this demonstration that he exclaimed, " Give me a fulcrum and I'll move the world."

The same temper that made so fertile an inventor as Archimedes despise the practical applications of his knowledge accounts also for the frivolous or destructive character of many of the machines invented at this period. The least frivolous are those associated with the name of Ctesibius. He invented a water-clock of great accuracy designed to indicate the hours of day and night under the system prevalent in classical antiquity, in accordance with which the lengths of the hours varied with the seasons of the year. He invented also a water-organ, a double-acting piston-pump, and a gun working by compressed air. The most interesting feature of these inventions is their dependence on the force exerted by air under pressure ; they are eloquent proof of the physical experiments now systematically pursued by the disciples of Strato of Lampsacus. The writings of Ctesibius have unfortunately perished. The earliest extant treatment of applied mechanics is a portion of a comprehensive work by Philo of Byzantium, who flourished either in the second or the third century B.C.

The techniques of the architects and engineers may never have found literary expression. If they did, the works have not survived. But one early and remarkable achievement in engineering has left its mark in literature, and archæology has lately confirmed the truth of the literary report. In the second half of the sixth century B.C. a Greek engineer, Eupalinus of Megara, presumably in the employ of the tyrant Polycrates, tunnelled the hill of Kastro in Samos to secure a supply of water for the town. The tunnel, which is over 900 yards in length, was begun from both sides of the hill, the points being determined by the existence of a brook at the side of the hill remote from the town and the necessity of an exit in the side of the hill near the town. How the problem of co-ordinating the operations of the two parties of diggers so that they should meet at the centre was solved is something of a mystery. They missed one another only by a few feet. A late Greek writer, Hero of Alexandria, who may not have written earlier than the second century A.D., has a problem which recalls this ancient feat of engineering. He states it thus : " To drive a straight tunnel through a mountain given the two starting-points." After giving the mathematical solution he confidently concludes : " If the operation is carried out in this way the two gangs will meet." But that a Greek engineer of Megara should have possessed this theorctical equipment in the sixth century is surprising. A celebrated Alexandrian engineering work, completed about 280 B.C., a few years after the death of the first Ptolemy, was the construction in the harbour of a gleaming tower of white marble, the Pharos, probably four or five storeys high, which served as a guide to ships by day. Its use as a lighthouse in the modern sense was, apparently, a later adaptation.

Of the three classical mathematicians of the Alexandrian Age one still remains to be discussed, Apollonius of Perga, who flourished about 240 B.C. His fame rests on his treatment of conic sections. The subject had been invented by Menæchmus, a pupil of Plato and Eudoxus, about 100 years before, and had been contributed to by Euclid and Archimedes. The task of Apollonius, like that of Euclid in the *Elements*, was to gather together, systematize, and complete the contributions of his predecessors. His work remained the classic on the subject down to Descartes. The work was originally in eight books, of which four are still extant in Greek and three more in an Arabic translation.

Finally, we have to speak of Hipparchus, a late flowering of the first Alexandrian period. He died about 125 B.C., and is generally considered the greatest of ancient astronomers. He was born at Nicæa in Bithynia—the Asiatic provenance of so many of the great scientists is a noteworthy fact— but lived most of his life at Rhodes and Alexandria. His name is connected with what, superficially considered, looks like one of the great retrograde steps in the history of science. He abandoned the heliocentric hypothesis of Aristarchus in favour of the geocentric view. But he had good warrant for his choice. Improved methods of observation brought more and more to light variation in the distances, and irregularities in the motions, of the sun, moon, and planets ; and the system of Aristarchus which made the earth revolve in a circle of which the sun was the exact centre failed to account for these phenomena. Hipparchus met the difficulty of accounting for the apparent motions of the planets by an elaborate system of epicycles which worked better than the theory of Aristarchus. His theory also had the advantage of not offending popular prejudice by removing the earth from its traditional place at the centre of the universe.

Hipparchus had at his disposal not only the results of some 150 years' work at the observatory of Alexandria, but also the much more ancient records of the Babylonian astronomers. We have already seen what use he made of these records in arriving at a more accurate determination of the lengths of the tropic year and mean lunar month (cf. p. 129). Two other remarkable achievements stand to his credit. He discovered the precession of the equinoxes, which is equivalent to the observation that the longitudes of all the fixed stars increase by fifty seconds per annum. His realization that this discovery was only made possible by the records handed down by previous astronomers inspired him to a prodigious effort in the interest of posterity. He mapped the positions of some 850 stars and noted their special characteristics so that later generations might be able to determine whether they changed in position, size, and brilliance with the course of time. Thus, at the close of its period of active growth as at the beginning, science was dominated by what is its characteristic morality, a sense of the solidarity of the generations, of the co-operation of past, present, and future. Art is long, life is short.

CHAPTER IX

THE GRÆCO-ROMAN WORLD

THE extraordinary prestige of the Greek language and culture in the Alexandrian Age is illustrated, among other things, by the fact that at this time representatives of older and different civilizations began to succumb to its power. An Egyptian priest Manetho and a Babylonian priest Berossus wrote histories of their own civilizations in Greek, fragments of which are still extant ; while a Babylonian astronomer adopted a Macedonian name, Seleucus, and published his scientific works in Greek. Alone in his age he supported the Aristarchan heliocentric hypothesis, and he is famous also for his correct theories about the motion of the tides.

Among the Alexandrian Jews also the Greek language spread, so much so that a demand arose for a Greek translation of the Hebrew scriptures for use in the synagogues of Egypt. It was an extraordinarily difficult task to express the religious and ethical ideas of the Chosen People, who had seen the hand of God in all their history, in the idiom that had been created by the scientists, philosophers, historians, and rhetoricians of Greek lands. The work, begun under Ptolemy II (284–246 B.C.), was completed in the course of the next 100 years. The Law and the Prophets were translated into the language of European culture ; and followers of Plato who felt with their master that there was little Greek prose that could be put into the hands of young people without exposing them to the danger of atheism were provided with a literature that answered their needs. A parallel was soon instituted between Plato and Moses, and a Wisdom was taught that did not offer itself as an argument addressed by one man to another but as a revelation given by the Creator to his creatures. A ferment was set up in men's minds which has continued unabated to the present day.

These were remarkable conquests of the Greek tongue in the east. For a time it seemed as if they were to be accompanied by an equally remarkable conquest in the west. At the end of the third century Rome, so powerful politically and so weak culturally, began to write her history in Greek. But this movement, initiated by Quintus Fabius Pictor and Lucius Cincius Alimentus, was checked in the first half of the second century by the arguments and example of old Cato, who scoffed at the notion of Romans writing Greek, and offered to Roman readers their own story in their own tongue. The possibility that Greek might become the sole cultural medium of Roman society in the west did not materialize, and from the middle of the second century European civilization gradually becomes bilingual. The Græco-Roman epoch had begun.

The period is not one of great advance in scientific knowledge. Such new knowledge as was won was the result of the efforts of Greek-speaking peoples, and with this we shall first deal. But the main task of the historian of science in this period is to estimate the degree of success achieved by the Romans, who originated nothing, in absorbing the science of the Greeks, for on this depended the type of culture which the Romans were able to disseminate among the newly conquered peoples of western Europe. And the truth unhappily is that in science the Romans were not apt scholars. To their failure it is largely due that it was only in the seventeenth century that western European towns attained the degree of scientific culture that had been achieved in Alexandria two thousand years before.

The progress that was made by the Greeks in the Græco-Roman period, a period which for our purpose may be considered to have lasted from the middle of the second century B.C. to the fall of the western Roman Empire in the beginning of the fifth century A.D., was chiefly in the sphere of mathematics and astronomy and the attendant subject geography. But the period also produced one great doctor and one great botanist.

The chief centre of scientific studies continued to be Alexandria, and here two new sciences, trigonometry and algebra, were invented. With regard to the former the chief facts are as follows. The astronomer Hipparchus had drawn up a table of chords of arcs in a circle subtending angles of different sizes. This is equivalent to a table of trigonometrical sines. The subject was further developed by Menelaus in the first century A.D.; then, in the second century, the great geographer and astronomer Ptolemy, to whom we shall have to return, prepared an elaborate table of chords subtending a series of angles beginning with ½ a degree and advancing by ½ a degree each time. Algebra was the creation of Diophantus, who flourished about A.D. 250. Of his *Arithmetica* in thirteen books the first six have survived. In these the science of algebra is virtually constituted, though without the convenient notation which, together with the name of the science, we owe to the Arabs.

One man, Ptolemy, to whom we have referred in the preceding paragraph, gave their definitive form both to the astronomy and geography of antiquity. In astronomy he took over the geocentric system as it had been left by Hipparchus and worked it out in a treatise generally known under its Arabic title of the *Almagest*. This remained the standard work on astronomy until the heliocentric hypothesis slowly won its way to acceptance in Europe after the publication of the work of Copernicus in 1543. The essential idea of the *Almagest* is that it accounts for the irregularities in the movements of the planets by the system of epicycles introduced by Hipparchus. That is to say, the planets are represented as moving in smaller circles each having its travelling centre on a large circle of its own, called the deferent. This was the final form of the answer given by the Greeks to the old problem raised by Plato ; and it is only fair to Ptolemy, that great scientist, to remark that he probably understood quite well that his answer was only a mathematical construction without claim to be an absolutely true representation of the motions of the heavenly bodies.

The geography of Ptolemy likewise resumes the tradition of Hipparchus. From Hipparchus he had inherited the doctrine that all geographical points should be determined astronomically, latitudes by reference to the elevation of the pole and longitudes by a method dependent on the observation of lunar eclipses. But the standard was too high a one for the age ; time and helpers were lacking for the performance of so great an undertaking. In Ptolemy's *Geography* six books out of the eight are devoted to fixing the position of some 8,000 places. They are located by latitudes north of the equator and longitudes east of the " Fortunate Isles." But the determinations are in the vast majority of cases not astronomically made, but derived from information supplied by travellers. Naturally they are often very erroneous.

In addition to his trigonometrical, astronomical, and geographical achievements, a pretty piece of experimentation stands also to the credit of Ptolemy. In his *Optics* he records the results of his investigation of the phenomenon of refraction. He experimented with various media—glass, water, and air ; and gives tables of the angles of refraction, not however always correct. His technique does not seem to have come up to the level of his original inspiration. It was no doubt phenomena observed in connection with astronomical research that suggested this line of investigation.

Of political as distinct from astronomical geography the chief exponent among the Greeks was Strabo. He was born in Amasea in Pontus about 64 B.C. of mixed Greek and Asiatic parentage, and having travelled much and written a successful historical work which has not survived, he devoted his energies and talents in the last decade of the pagan era to the composition of a panoramic review of all the lands within the borders of the Roman Empire, which has come down to us practically intact in all its seventeen books. His first book, in the now common fashion, offers a critical review of the work of his predecessors in the geographical field. In his second book he attempts, without too much success, to handle the mathematical part of geography. In the remaining fifteen books he has reached the aspect of his subject in which he is most at home and gives in fluent readable Greek an intelligent account of the manners, institutions, and history of the chief countries of the civilized world. The work is typical of the age. A masterpiece of Greek scholarship, it is the creation of a man of mixed descent born on the remote shores of the Black Sea, who is a strong supporter of the political dominance of Rome.

Equally typical of the new Græco-Roman world is the career of the one great doctor produced between the time of Herophilus and Erasistratus and the rebirth of medicine at the Renaissance. Galen was born about A.D. 130 in Pergamum, at that time a seat of learning hardly less important than Alexandria itself. Having studied philosophy and medicine in the town of his birth, he continued his medical studies at Smyrna, Corinth, and Alexandria, and eventually made his way to Rome where he was court physician to the emperor Marcus Aurelius. He was one of the most voluminous writers of antiquity; though more than half his works are lost, yet about a hundred treatises, medical and philosophical, have survived. So voluminous a writer is bound to be careless and prolix and to invite oblivion for much of his output; and such is the character and such has been the fate of many of Galen's works. Nevertheless both as an author and researcher he is of the first importance.

Basing himself on the tradition of the Hippocratic school, he studied and commented upon their writings, and supplemented their theory by the teleological teaching of Aristotle. Like Aristotle he subordinated structure to function, and had as his ideal the ultimate revelation of the divine mind in every particle of nature. But, again like Aristotle, he was an active researcher, practised in the dissection of dead and living animals, and he attained results of permanent value. The most remarkable were in neurology. His experiments on the spinal cord, described in his treatise *On Anatomical Operations*—he severed the spinal cord of a monkey at various points and showed how the powers of motion and sense in the parts below the section are destroyed—are a classic page in the history of science, and had direct influence on Vesalius in the sixteenth century. His neat demonstration also that the arteries contain blood and not air, by which he corrected the mistaken view of Erasistratus, form an important step in the ascent to the discovery of the circulation of the blood. The experiment consisted in ligaturing an artery firmly at two points some distance apart. On incision at a point between the ligatures blood was found to fill the artery. The result and the technique were both passed on to Harvey, the discoverer of the circulation of the blood, through the intermediacy of Vesalius and his disciples.

Owing to the mastery with which he handled every aspect of the art of medicine, anatomical, physiological, therapeutical, ethical, philosophical, and historical, the writings of Galen remained until the sixteenth century the unquestioned authority in heir sphere. This position of authority they well deserved; that the authority was unquestioned was not Galen's fault. And when Vesalius, the restorer of anatomical research in the sixteenth century, ventured to point to mistakes in Galen, it was with the followers rather than with the master that his quarrel lay. By reviving the practice of Galen he routed the Galenists.

One other masterpiece of science produced in this age held the field till modern times. This was the *De Materia Medica* of Dioscorides in five books, which consists of a catalogue and description of some 600 plants supposed useful for medicinal purposes. The author, like all the scientific intellects of the time, came from the eastern part of the empire. He was born in Cilicia. His book was written in the middle of the first century A.D.

When we turn to consider the progress of the Latin-speaking portion of the empire in assimilating Greek scientific culture, the first and greatest success to be recorded is the re-creation of the Latin language itself. As an instrument for deliberation and administration the Latin tongue already in the middle of the third century B.C. had some claim on the admiration of mankind. But it was a stiff inflexible medium with an impoverished vocabulary suited only to the expression of a narrow range of political ideas. The early historians of Rome were either Greeks, or Romans who wrote in Greek, and Greek throughout the whole of our period remained the chief medium of higher education in Rome. But Roman self-respect demanded that Roman ideals should not find their only expression in the language of the conquered Greeks. A movement arose for the establishment of a national literature in the Latin tongue. The Latin writers went to school to the Greeks. Generation after generation they wrestled with the problems of translating, adapting, or imitating the masterpieces of Greek literature in their inadequate medium. The vocabulary was extended, turns of expression were multiplied, and the range of ideas expressible in the language enormously increased. Every Roman writer was a conscious grammarian and a conscious stylist. The whole science of Greek grammar was taken over and adapted, sometimes a little awkwardly, to the Latin tongue ; while the study of Greek rhetoric gave balance and subtlety to a medium that soon aspired to the heights of eloquence. By the end of the republican period, and still more after the Augustan Age, that is to say roughly about the commencement of the Christian era, Latin literature possessed a series of masterpieces the ability to understand and appreciate which still stamps a man with the narrow, but not meaningless, title of scholar, and is a useful guarantee of a capacity for high literary culture. Latin was moulded in turn to the purposes of the poet, the historian, the orator, and the philosopher. Up to quite modern times it bore the main stress of western European endeavour in these fields. Even in the sphere of science the first masterpieces of modern speculation and research were composed in this medium and the use of Latin for scientific purposes continued, with increasing competition from the modern local vernaculars, up to the eighteenth century.

The first important achievement of the Latin tongue in incorporating the philosophical and scientific ideas of the Greeks was the epic *De Rerum Natura* of Lucretius (died 54 B.C.). It is also the greatest. The atomic system of Democritus had been given a new vogue in the third century B.C. in Athens by the philosopher Epicurus. Epicurus represented a trend of thought diametrically opposed to that of Plato. While Plato put all the emphasis on the next life, Epicurus believed only in this. While Plato warred against the scientific materialists, Epicurus based his philosophy upon them, rejecting only the theory of mechanical determinism. For Epicurus peace of mind lay in the realization that the destructive phenomena of nature, thunder and lightning, earthquakes and inundations, plagues and pestilences could all be explained by the action of atoms in the void and did not imply the hostility to men of angry gods ; and in the further realization that the soul, likewise a structure of atom and void, is mortal and therefore not subject to the tortures in a future existence which popular imagination and Platonic prose had presented with such terrifying vividness.   It was the doctrine of Epicurus, with its basis in the atomic system, that Lucretius took as his theme, and he cast it into epic form on the model of the philosophical poem of Empedocles.   His poem contains nothing original except the fervent and noble eloquence of the writer, and his eminent capacity for the organization and orderly exposition of his material.   It is admittedly a masterpiece of literature, the world's supreme philosophical epic ; but from a certain angle it is also a masterpiece of scientific thought, if science is not only a technique but a philosophy, a mentality, a way of looking at things, a faith in reason.   The holy delight in the spectacle of nature and in the knowledge of her laws, the necessity of a knowledge of these laws in order to live rightly, the duty to submit the mind to the evidence of the observed fact—these ideas have never elsewhere been expressed with such power and beauty as in the stern eloquence of the *De Rerum Natura*.

Contemporaneously with Lucretius, Cicero made a remarkable effort to incorporate in the literature of Rome what he regarded as best in Greek thought. A superb and facile translator, he turned into competent verse in his youth a Greek astronomical poem, the *Phænomena* of Aratus, in which the astronomical ideas of Eudoxus were expounded in popular form. But his chief activity was in prose. Altogether less serious than Lucretius, and unable in the intervals of a busy public life to spare time for such strenuous study as went to the composition of the poem *On the Nature of Things*, he contented himself with a more popular and less exacting theme and produced a work, in three books, drawn likewise from Greek sources, *On the Nature of the Gods*. In it he expounded the Epicurean and Stoic views of the government of the universe and criticized them from the sceptical standpoint of the later Academy. He was an acute thinker and exquisite writer, with a lively impressionable mind, and a capacity for moral elevation which found relief in literature after the sordid exigencies of Roman public life.

One of the most courageous and useful of his treatises is that *On Divination*. In it he puts to searching analysis the whole question, so important for ancient society, of the possibility of forecasting the future by signs and omens, by the stars, by dreams, and so on. " There is an ancient belief," he tells us in his opening words, " handed down to us even from mythical times and firmly established by the general agreement of the Roman people and of all nations, that divination of some kind exists among men." Against this view, established in the opinion and practice of his own people and state, he argues with point and vigour, and concludes on a note of uncompromising condemnation. " Speaking frankly, superstition, which is widespread among the nations, has taken advantage of human weakness to cast its spell over the mind of almost every man. The same view was stated in my treatise *On the Nature of the Gods* ; and to prove the correctness of that view has been the chief aim of the present discussion. For I thought I should be rendering a great service both to myself and my country if I could tear this superstition up by the roots." It is a noble page in Latin literature.

While Lucretius and Cicero approached science from the point of view of its power to liberate the mind from superstition, a later writer, Celsus, (about A.D. 30), is our best example among the Romans of a purely scientific author. He undertook the composition of an encyclopædia of various branches of Greek science. Unhappily, with the exception of the section dealing with medicine, it has wholly disappeared. The work *On Medicine* is of great importance. Like the writings of Lucretius and Cicero it is drawn wholly from Greek sources and shows no evidence of original work in research. But like them also Celsus shows the Roman capacity for mastering his material and giving it logical and coherent expression. His treatise is on the whole the best general treatment of its subject bequeathed to us by antiquity.

The same mastery of his material is not claimed for Vitruvius, a military engineer of the time of Julius Cæsar and Augustus, who being pensioned off in old age devoted his leisure to the composition of a work *On Architecture* in ten books. The first seven books deal with architecture proper, the eighth with water and water-ways, the ninth with water-clocks, the tenth with machines. The work was of practical use at the Renaissance and is studied now for its historical importance. But its treatment of its subject is often obscure and difficult, raising the question whether Vitruvius was fully capable of understanding his Greek originals.

The Roman mind ran to erudition rather than to investigation, and a characteristic product of it was the encyclopædia. One encyclopædist we have already met in the person of Celsus. Another was Varro (116–27 B.C.), the most learned of the Romans. He distinguished nine " liberal arts "—grammar, dialectic, rhetoric, geometry, arithmetic, astronomy, music, medicine, and architecture—and wrote upon all of them. Two only of his writings have been preserved, a treatise *On Farming* and a portion of his work *On the Latin Language*.

A later encyclopædist, the elder Pliny (A.D. 23–79), was more fortunate in that his work has survived. He is the author of a *Natural History* in thirty-seven books, dealing with cosmology, geography, anthropology, zoology, botany, medicine, mineralogy, magic, and art. It is, by modern standards, an ill-arranged miscellany without much guiding thread ; but such industry went to its compilation that there is justification for the statement of Lynn Thorndike in his *History of Magic and Experimental Science* that it " is perhaps the most important single source extant for the history of ancient civilization." Pliny has nothing of his own to contribute to science except his omnivorous curiosity into all Nature's secrets, in which he feels himself to be unique among the Romans ; but as a reporter of other men's discoveries he has deserved well of posterity. His book is compiled from some 2,000 works by nearly 500 authors, of whom more than two-thirds were Greek. His nephew, Pliny the Younger, describes it as " a work remarkable for its comprehensiveness and erudition " (which is no more than the simple truth), " and not less varied than Nature herself " (which may pass as better deserved than many similar compliments).

To Pliny, and to a lesser degree also to Vitruvius, we are indebted for much of what we know of the

chemistry of the ancient world. As has been implied in our first chapter, the history of chemistry
among the Egyptians and Babylonians must be
inferred mainly from analysis of the surviving
objects of their industries. It is generally admitted
that the Greeks derived their chemical arts from the
Egyptians, but the earliest written records from
Egyptian soil are MSS written in Greek of the
third century A.D. These MSS give some hundreds
of chemical recipes, but the problem is to date the
origin and development of this obviously traditional
knowledge. At a date some six hundred years
earlier we have interesting information of the state
of Greek chemical knowledge in the writings of
Theophrastus, the pupil of Aristotle. In his work
On Stones he describes many natural minerals and
their products in a recognizable way. We may
note particularly his clear account of Plaster of
Paris. From his *History of Plants* may be instanced
his description of the methods of preparing char-
coal and of recovering pitch from resinous trees.
And in his work *On Scents* he gives us the first
mention of the water-bath as a device for securing
gentle heat in the preparation of perfumes.

But our knowledge of ancient chemistry would
be much diminished without what Vitruvius and
Pliny have preserved. Both, for example, describe
the process of recovering gold in small quantities
by amalgamation with quicksilver. And Pliny's

account of the manufacture of semi-precious stones, such as beryl, carbuncle, sapphire and opal, is well known. There are no frauds, he says, which bring greater profits ; and for the benefit of possible dupes he mentions tests that may be applied, namely examination in bright sunlight, relative weights, the feeling of coolness in the mouth, and differences in hardness. Examples might be multiplied indefinitely to illustrate the range and variety of the chemical knowledge of the ancients. An adequate theoretical basis was, however, not found. The atomic theory of the constitution of matter was neglected, and the traditional four elements, earth, air, fire, and water afforded an impossible basis for chemical analysis.

Seneca, the Stoic philosopher and tutor of the emperor Nero, also wrote on scientific subjects in his work *Quæstiones Naturales*. He always writes with lucidity and point, and discourses popular science in a pleasing style on such topics as the rainbow and the reasons for the periodic rising of the Nile. But his work belongs, in a sense, to a pre-scientific age. Celsus, in his work on medicine already referred to, has a famous phrase about Hippocrates, to the effect that " he was the first man to separate medicine from philosophy." The distinction between science and philosophy Seneca never perceived. For him scientific truth is established by argument rather than by experiment. The most that can be said for him is that, like other enlightened Romans, he was interested in the results if not in the method of scientific pursuits.

The only Roman geographer was Pomponius Mela, who was born in Spain and flourished in the middle of the first century A.D. It is hardly necessary to say that his work is of the descriptive rather than of the mathematical kind. Drawing on somewhat antiquated Greek sources he describes a circular tour of the Mediterranean lands. The style, which is far from unpleasing, is characteristic of Roman scientific writing. It is neither that of the text-book nor of an original enquirer, but of an artist who works up borrowed material into literary form. Mela lays down the general plan of the earth as known to him, the disposition of the main masses of land and water, and the division of the earth into the five zones. Then he deals with the three known continents, Africa, Asia, and Europe. He gives no distances and no measurements. The work, therefore, hardly comes up to the level of the best Roman reproductions of Greek science. The treatises of Celsus and Vitruvius, defective as the latter is, were intended as solid practical manuals. Mela's work is more of the kind that French popularizers used to describe as " for ladies," meaning that the difficult detail was omitted. It is unfortunately only too true that much Roman scientific writing consists in transforming the solid science of the Greeks into Latin *belles lettres.*

One notable failure of Roman culture to assimilate Greek science remains to be mentioned—mathematics. The Roman attitude is illustrated by Cicero in the opening of his *Tusculan Disputations.* " The Greeks," he remarks, " held the geometer in the highest honour, accordingly nothing made more brilliant progress among them than mathematics. But we have established as the limit of this art its usefulness in measuring and counting." Nor must it be supposed that Cicero is here criticizing the Roman attitude. At least, if such is his intention, he is careful not to avow it openly, for he has just given it as his opinion that " the Romans always showed more wisdom than the Greeks in their inventions, or else improved what they borrowed from them, supposing they thought it worthy of serious attention." Apparently mathematics did not fall within this category. There is a sense in which one may legitimately sympathize with the Romans. The frequent insistence by the Greeks on the uselessness of mathematics for practical ends naturally acted as a deterrent with the practical-minded Romans. Nevertheless, the consequences for Roman civilization of their inability or refusal to learn mathematics were serious. They borrowed the results of Greek science without acquiring the method, taking over the enunciations of Euclid's propositions without troubling with the proofs. Consequently they remained dependent on the intellectual resources of the eastern half of the empire. The survey under Augustus was carried out by the aid of specialists from Alexandria, just as Julius Cæsar had relied on the aid of an Alexandrian astronomer, Sosigenes, for his reform of the calendar. But when the eastern half of the empire was sundered from the west, when the knowledge of Greek declined in the western half and Latin remained as the one medium of education, its scanty content of positive knowledge, its insecure grip on the meaning and the method of science, and its failure ever to have established a tradition of research, eliminated the possibility of an education based on a knowledge of nature or even adequately salted with it. Education was directed to words rather than to things. The verbal disciplines, rhetoric, grammar, and logic, did not wholly perish ; but nature was a book that was not directly studied.

CHAPTER X

SUMMARY AND CONCLUSION.   THE DECLINE OF
ANCIENT SCIENCE

In the middle of the second century B.C. the rate
of advance in Greek science noticeably slowed
down.   By the end of the second century A.D.
progress had stopped.   In the centuries that followed, until the Arabs took up the running, science
was in full retreat.   Before attempting to describe
the nature of the decline a brief résumé of our
narrative will be of help.

The continuous history of science begins with the
Ionian Greeks in the sixth century B.C.   The
Greeks, however, had predecessors ; and though
the connection between the older civilizations and
Greece is far from being understood, recent research
tends to bear out the truth of the old Greek tradition that their science had its roots in the ancient
civilizations of Egypt and Babylon.

In both these centres we have evidence of a
multitude of techniques which imply empirical
acquaintance with the properties of a great many
natural objects and a rational use of this knowledge.
This is the raw material of science, and it is not

221

impossible that further archæological discoveries may contribute to give this technical knowledge a more scientific character.

In addition to their techniques we have evidence from both civilizations of the beginnings of genuine mathematical science. It never disengaged itself from practical applications and never became organized as a logical series of deductions from a few self-evident principles ; but the germ of the idea of proof is present, and one feels compelled to accord the body of rules a higher name than that of a mere technique.

In both civilizations also are to be found remarkable achievements in the acquisition of positive knowledge in limited spheres. Egypt presents us with an astonishing fragment of genuine medical science, which raises the question whether there may not be more, or once have been more, where this came from. We are reminded that we cannot in any sense write a history of science in the ancient east ; the material is too fragmentary. In Babylon, on the other hand, astronomy reached remarkable heights. Observations gathered over centuries and interpreted mathematically established a genuine body of knowledge that was continually tested in practice, refined and improved on. It is no argument against the scientific character of this knowledge that the motive of its creation was a mistaken belief in the terrestrial influence of the heavenly bodies. If this criterion were to be rigorously applied there would be little science in the world, for it is all mixed with error and pursued for motives that the progress of time has shown or may show to be illusory.

When we pass from the orient to Greece, we find that the novelty of the cosmology of the old Ionian school is that it rested on the implicit assumption that the universe as a whole is intelligible in terms of everyday experience. The fragments of their writings that have come down to us still impress by their optimistic naturalism and confident rationalism. It is as if a new spirit had come into the world.

Their optimism, however, was naïve. The problem is more complex than they supposed. " The difference between reasoning and the facts on which it is based " (to steal a phrase from M. Arnold Reymond) impressed itself upon men's minds, and the way was prepared for the great revolution in outlook associated with the names of Socrates and Plato. The conditions of scientific knowledge became the subject of searching analysis. Man's universe fell apart into the mental and the material ; and after a period in which Plato had tried to define science as a pure activity of the mind, Aristotle by the constitution of the sciences of logic and psychology provisionally bridged the gap. From the Platonic system, however, has survived the attitude which regards physical science as dangerous, as if the knowledge of nature were an obstacle to the knowledge of God.

The result of the Aristotelian analysis of existence was to find in it the two elements, matter and form. Form is the intelligible aspect of what exists, matter the tangible. The object of science is to arrive at a knowledge of forms. The doctrine of form applied to the world of plants and animals led to valuable results. Plants and animals present the spectacle of a process, a life-cycle, in which from relatively small and chaotic beginnings the perfect individual matures, and then declines and dies. If we regard the individual at its maturity as having fully realized its form, we can understand the life process as a continual imposition of form on matter. Form is that which, both in the individual and the species, imposes unity and identity on a changing material content. The biological sciences were constituted on this basis in the Lyceum.

The doctrine of form was not so helpful in the sphere of physics. Physical processes do not culminate in the maturity of an individual's form which can be looked upon as the goal of nature's endeavour. It is a question rather of connecting event with event and determining the law of their connection. What was wanted for the advance of physics was a doctrine of force rather than a doctrine of form. Neither the teleological conception of Aristotelian biology nor the standard of logical intelligibility demanded by the mathematical sciences was helpful to the advance of a science in which progress was to be made later at the Renaissance only by the method of observation and experiment and a willingness to accept the fact *that* one event follows on another without seeking to know *why* it does so.

Only one school, the Hippocratic doctors, approached the realization of this point of view. Their movement is also responsible for the first clear enunciation of the ideal of science as the servant of humanity. If the idea is now common in men's minds (however absent from their practice) that science ought to benefit the human race, the first to proclaim this ideal and to organize themselves to give it effect were the doctors of the Hippocratic school.

The science that achieved its ripest maturity with the Greeks was mathematics, which to a large extent set the standard for all the other sciences. Their ambition to represent each department of mathematics as a logical series of deductions from a few self-evident principles revolutionized the science as they had taken it over from the Egyptians and Babylonians and was highly successful so long as the science remained pure and unapplied. But when it was sought to organize observational sciences on the same deductive plan the result was not so good. The statics and hydrostatics of Archimedes are set forth in a series of beautifully connected propositions whose logical coherence excites the admiration of the reader. But the spirit is very different from that of a modern laboratory where one is taught the technique of experiment, weighs, measures, tabulates results, and strikes an average. Archimedes is still under the domination of the idea, true enough in the abstract world of pure mathematics, that what is logical *is*. The notion of a scientific law as an average would have been quite strange to him. It is hardly surprising that physics and chemistry made little secure progress. The Greeks experimented with steam and compressed air ; they investigated the laws of refraction ; they meditated the problem of the transformation of substances into one another ; and they had a theory of atomism ready to hand. But they never made a systematic practice of measuring physical events. In general the connection between the mathematical and the physical eludes them.

By the dawn of the third century B.C., with the commencement of the Alexandrian Age, the renown of science was such that practical governors saw in it an instrument of power and made it part of their policy to promote its progress. Plato throughout a long life had preached the necessity of enthroning wisdom in the seat of power, and his Academy had fought valiantly to this end. But the unpractical nature of the studies there pursued largely negatived the efficacy of his plea. Now, at Alexandria, the resources of a powerful state were for the first time put behind the scientists, and where advance was possible it was made with astonishing rapidity.

The great practical achievements of the Alexandrians were the measurement of space and the measurement of time. They mapped the heavens and the earth; they brought the calendar to a marvellous accuracy; they improved the sundial and the water-clock to such a degree that the Roman empire became a clock-regulated society. But when we search into the causes for the slowing down of the advance of ancient science we must conclude that essentially it was an internal phenomenon. Greek science was not killed, it died. It had reached the limit of possible expansion within the mould in which it was cast.

It is a different question why the theoretical knowledge available was not turned to the invention of machines for the production of commodities, or why the machines that were invented were of the nature of toys rather than labour-saving devices. Here the sufficient answer seems to be that the prevalence of slavery and the cheapness of slaves at the time when the necessary knowledge was available robbed society of the motive for this development. It seems incredible, for instance, that machinery for mining would not have been invented, were it not that the labour of slaves and criminals was so cheap. Yet sheer human stupidity may account for much. In the matter of transport, to choose another example, a recent enquiry has revealed the astonishing fact that throughout antiquity the method of harnessing draft animals was so inefficient that animal traction was inadequate to the task of moving the heavier loads. They were moved by men. A lucky victory involving the enslavement of the conquered army would be the occasion for extensive quarrying, the shifting of huge blocks of stone, and the erection of temples to the gods. Piracy, as the permanent source of the slave-supply, was winked at by the Roman government. It was an integral part of the social system.

The aversion to manual labour on the part of free men must have operated also to the detriment of science. The experimental scientist cannot dispense with his hands. The great advances in biology made by Aristotle, and in human anatomy and physiology by Herophilus and Erasistratus and Galen, were made at the cost of severe and disagreeable personal toil in dissecting dead and living animals. No serious advance in the knowledge of the structure and functioning of the human body was made between Galen and Vesalius, and Vesalius is emphatic in ascribing this to the dropping of the practice of personal dissection owing to the contempt of freeborn men for manual tasks. " It was when the more fashionable doctors in Italy, in imitation of the old Romans, despising the work of the hand, began to delegate to slaves the manual attentions they deemed necessary for their patients . . . that the art of medicine went to ruin."—" When the whole conduct of manual operations was entrusted to barbers, not only did physicians lose the true knowledge of the viscera, but the practice of dissection soon died out, doubtless for the reason that the doctors did not attempt to operate, while those to whom the manual skill was resigned were too ignorant to read the writings of the teachers of anatomy." Over and over again he repeats this charge, exhorting the young men to return to the practice of the ancient Greeks and asserting that his own achievement in restoring anatomy to its rightful place in the art of medicine has been wholly due to his personal practice of dissection.

We may find, then, in the limitations of the Greek ideal of science, in the inaptitude of the Romans for this study, and in the social conditions of the time, good and sufficient reason for the slowing down of the rate of progress in science and the failure to apply the theoretical knowledge already won to practical ends. But it still remains to ask why the gains that were made should have been lost. From the fifth to the tenth century, though in a lesser degree in the Greek-speaking east than in the Latin west, humanity went backward. The bulk of the record of ancient science actually perished ; the rest mouldered neglected in monastic libraries, without whose shelter it would have disappeared, but it perished effectively from men's minds. It is into this phenomenon we now have to enquire.

A fashion has established itself of ascribing the decay of ancient science to the influence of Christianity. The early Church was indeed ignorant and where it prevailed the heritage of science was not safe. But Christianity was only one of a number of competing religions which were hostile or indifferent to the pursuit of positive knowledge, and it seems more in accordance with the truth to see in the decay of science one of the conditions for the spread of these religions than to see in the triumph of one of them the explanation of the decay of science. But the question is a vast and complex one, and no more can be attempted here than to suggest by a few observations and illustrate by a few striking examples the nature of the revolt from science.

Three characteristics of Greek science may first be stressed. Its gaze was directed to this world. Its method of propagating itself was by an appeal to reason. Its urge to transcend the limitations of mortality was satisfied by the ideal of service to posterity. It was therefore the product of a society looking to a future of progressive well-being for humanity on its earthly home.

The religions, and to some extent Platonism, directed man's gaze to another world ; sought rather to make their converts undergo a transforming experience than learn something ; and taught that the true business of life was the preparation of the soul for an immortality of bliss in some other sphere.

The most important of these religions were the worship of Cybele, the mother of the Gods, of Attis, of Bacchus, of Isis and Osiris, and of Mithra. All these cults had won numerous adherents throughout the Græco-Roman world before the triumph of Christianity. Their principal difference from Christianity was that they worshipped deified powers of nature, while Christians worshipped a God who stood outside nature. All alike responded to a need different from that which had called science into existence. The religious adept sought assurance of personal salvation, escape from the burden of guilt, consciousness of communion with God ; and fastings, vigils, ceremonies of initiation, sometimes mutilations, were paths to his goal.

The mystical and the scientific outlook have both contributed to the complex structure of the modern mind. There is probably no man who is not susceptible in some degree to the appeal of both. It is true that people who know that water is H_2O rarely join quite simply with St. Francis in

thanking God for water " because it is so clean " ;
but they are stirred when Keats calls attention to—

> The moving waters at their priestlike task
> Of pure ablution round earth's human shores.

There is a wide range of human experience with
which science seems to have nothing to do. The
European mind would be impoverished if the
tradition of the mystery religions were eliminated
from it. It is out of this tradition Crashaw speaks
in his admired apostrophe to St. Theresa :

> O thou undaunted daughter of desires !
> By all thy dower of lights and fires ;
> By all the eagle in thee, all the dove ;
> By all thy lives and deaths of love ;
> By thy large draughts of intellectual day,
> And by thy thirsts of love more large than they ;
> By all thy brim-fill'd bowls of fierce desire,
> By thy last morning's draught of liquid fire ;
> By the full kingdom of that final kiss
> That seiz'd thy parting soul and seal'd thee His ;
> By all the Heav'n thou hast in Him
> (Fair sister of the seraphim !)
> By all of Him we have in thee ;
> Leave nothing of myself in me.
> Let me so read thy life, that I
> Unto all life of mine may die.

The trouble is that the thirsts of love of which the
poet speaks seem sometimes to find no satisfaction
in draughts of intellectual day. History seems to
show that obscurantism and persecution are among
the handmaids of religion.

While the mystery religions were absorbing men's minds in the question of their individual salvation and causing positive knowledge to seem of little account, another idea, which on the whole militated against a scientific outlook, was capturing adherents everywhere. This was the astrological view of the government of the universe. We give a short account of the system from the statement of the historian Diodorus Siculus. The Chaldeans teach, he tells us, that the cosmos is eternal, that it has had no beginning and will have no end. It is an orderly whole under the control of divine providence. None of all the things in heaven happens at random or of itself, but all are accomplished in accordance with a definite and established decision of the gods. Having observed the stars for very long and learned the motions and the powers of each, the Chaldeans can foretell much of the future to man. Special attention is due to the five planets, which are called Interpreters, because by their varying paths among the fixed stars they reveal the mind of the gods to men. Under their sway are thirty stars called Counsellor gods, with duties of supervision over heaven and earth. Of these gods twelve have special authority, and to each of them is assigned one of the months and one of the signs of the zodiac. The motions of the sun, moon and planets have decisive influence for good or evil on men at their birth. There are also twenty-four stars called Judges, who judge the living and the dead.

In general outline this teaching harmonizes fairly well with the astronomical religion inculcated by Plato in his later years. It was fully adopted by Stoicism, which became the characteristic philosophy of Rome. It has some solid advantages. In spite of its superstitious side it was, owing to the astronomy it incorporated, the most scientific religion of antiquity ; and by its substitution of the celestial luminaries as objects of worship in place of the local deities that presided over separate communities it prepared the way for the advent of a universal religion and fitted in with the centralizing policy of a great empire. But it enthroned in the minds of millions a view of the universe that made the old philosophers of Ionia with their cosmological theories appear presumptuous atheists ; and it encouraged a belief in divination by the stars against which enlightened intellects like Cicero battled in vain.

A further misfortune was that this astrological cosmology invaded the body of man. We have seen above that the positions of the sun, moon and stars were supposed to have decisive influence on men at their birth. A sympathy was imagined to exist between the parts of the human body and the signs of the zodiac. The human body was divided into twelve parts each of which was related to a zodiacal sign, and influence was supposed to be exerted on a man by the planets, the moon and the sun according to the sign in which they stood at his birth. When we speak of a man as having a jovial or mercurial temperament, describe him as having a saturnine disposition, or call him a lunatic, we are still speaking the language of the astrological view of human character and destiny.

This scheme, known in the Middle Ages as the doctrine of Macrocosm and Microcosm, could claim the countenance of Plato in the *Timæus*. By ill-luck this fantastic dialogue, part of which had been translated by Cicero, and another version of which was made in the fourth century by a Christian writer Chalcidius, represented in the Dark Ages the high-water mark of Greek natural philosophy. Its support of the doctrine of Macrocosm and Microcosm was a misfortune ; for the doctrine operated to the complete destruction of the Hippocratic and Galenic tradition of medicine.

The translation of the Hebrew Bible also brought with it influences that militated against the scientific outlook. Of these the most deadly was the doctrine of the creation of the world from nothing. Plato in the *Timæus* had taught a doctrine of creation. The probable interpretation of his words, however, is not that God created the world out of nothing, but simply that he had brought order into a pre-existing chaos. This may also have been the original Hebrew view (the account in *Genesis* is open to either interpretation), but in the generations immediately preceding the birth of Christianity Hebrew literature speaks more decidedly in the sense of a creation from nothing. In the second book of *Maccabees* it is stated explicitly that God made the heavens and the earth out of nothing.

With the acceptance of the literal accuracy of the creation story in Genesis the door was opened for a strange development. By a curious process of logic it was argued that since God made the world in six days, and since a thousand years was but a day in the sight of the Lord, the total duration of the world would be but 6,000 years. After that would come the Millennium, corresponding to the Sabbath on which God rested, when for 1,000 years the Saints would reign on earth. Then there would come the Judgment, and heaven and earth would pass away.

This theory of the six thousand years' limit to world history, originating with a Syrian thinker Bardesanes, had a tremendous vogue. It fitted in well with millennial aspirations that even in pre-Christian times had been associated with the astrological view of the universe. According to a widely-held view which goes back to the remote beginnings of Babylonian culture, the sun was looked upon as the dispenser of justice, and it was imagined that there was a model City of the Sun in heaven which in God's good time would be imitated on earth. Armies of revolted slaves, hoping for a new order of society, enrolled themselves under the banner of the sun. " Thy will be done on earth even as it is in heaven " represents fairly the aspiration of these believers in the City of the Sun.

But the limiting of the course of this wicked world to six thousand years, comforting as it must have been to suffering millions, was a fatal blow to the presuppositions of science. In the fourth century a Latin father, Quintus Julius Hilarianus, distressed by the fact that some of the brethren thought the world was more than 20,000 years old, worked out a scheme of chronology that dominated the Middle Ages and lasted down to quite modern times. The fixed points in this scheme were the Creation or the Beginning of Time, the Nativity or the Fullness of Time, and the Judgment or the End of Time. In the precise chronology worked out by Hilarianus God had created the world at six o'clock in the morning of the spring equinox 5,899 years before. According to calculations based on Biblical history, from the Creation to the Flood was 2,257 years, from then to the Exodus 1,389 years, from then to the Fall of Jerusalem and the Captivity 1,168 years, from then to the Passion 716 years, and 101 years were still left before the Millennium.

This dating would already make nonsense of such

ideas as that of Herodotus that it might take ten or
twenty thousand years for the alluvial deposit from a
river to create a territory like that of the Nile
delta, or the idea expressed by Plato that 9,000
years before his time Attica had been a fertile
country deep in soil which slow erosion had carried
away. Still further complications were produced
when the Millennium failed to come. Every year
that the world lasted after the 101 allowed to it by
Hilarianus had to be taken off the beginning. Thus
while Hilarianus had allowed 5,550 years between
the Creation and the Nativity, Bishop Ussher in the
seventeenth century was compelled to reduce the
interval to 3,999 years, 2 months, 4 days and 6 hours,
which still in the seventeenth century allowed some
time before the end of the world. It did not,
however, leave enough time for the previous history
of mankind.

Unhappily this was no longer of much account
for history had in any case been largely forgotten.
This had been the inevitable result of the literal
acceptance of the inspiration of the Bible. The
ancient Hebrew literature is a very precious heritage
of humanity, but it has its limitations. The Jews
were late-comers on the scene compared with the
Egyptians and Babylonians, and had suffered much
from both. What they record of Egyptian and
Babylonian history is naturally insignificant and
written from the angle of an oppressed race. If,

then, the Biblical account of world history be accepted at its face value, the believer is shut out from a full and proper knowledge of the great civilizations of Egypt and Babylon. Such was the case for nearly two thousand years. The archæologist is only now painfully recovering the knowledge of the civilizations of the near east.

As with history, so with other branches of knowledge. The effort to interpret the Bible literally continually produced strange results. The most learned of the Greek fathers, Origen, in the second century, endeavouring to make room for all the animals in Noah's ark, had come to remarkable conclusions on its size. He estimated it at 90,000 cubits long, 2,500 cubits wide, and 900 cubits high. This makes of Noah the greatest nautical engineer in history. But as the ark expanded the world shrunk. St Augustine, in the beginning of the fifth century, was constrained to deny the existence of men at the antipodes, because, being inaccessible to the Gospel message, they would be cut off from the hope of salvation. And a century later Cosmas Indicopleustes, in his *Christian Topography*, set out to refute the theory that the earth is round. His aim was to prove a view, then, alas, of some two hundred years standing in the Church, that the tabernacle of Moses was a model of the universe. Basing himself on irrefragable scriptural authority he showed that the earth must be a flat plain with high walls north, south, east and west, over which closes a semi-cylindrical lid, the sky. It has a high mountain towards the north round which the heavenly bodies are carried by angels, thus producing the phenomena of night and day, eclipses and so forth. Shelley wrote with more literal exactitude than he perhaps knew when he complained that the advent of Christianity had made the world " an indistinguishable heap."

In truth physical science is not encouraged by the Bible. The essential of the narratives in both Old and New Testaments is miraculous ; and Jehovah definitely discourages natural philosophy. The list of questions with which He seeks to silence Job is significant. " Then the Lord answered Job out of the whirlwind and said : Who is this that darkeneth counsel by words without knowledge ? Gird up now thy loins like a man ; for I will demand of thee and answer thou me. Where wast thou when I laid the foundations of the earth ? Whereupon are the foundations thereof fastened ? Who shut up the sea with doors ? Hath the rain a father ? or who hath begotten the drops of dew ? Canst thou bind the sweet influences of the Pleiades, or loose the bands of Orion ? " Job, of course, understood perfectly what was expected of him. " Then Job answered the Lord and said : Behold I am vile ; what shall I answer thee ? I will lay my hand upon my mouth."

Among the Christian writers a regular practice developed of confounding Greek physicists in the manner Jehovah had found so successful with Job. The gentle author of the *Epistle to Diognetus*, a work of the second century, in which the more estimable sides of Christianity, its charity, its meekness, its inner assurance of peace, are strongly present, falls a victim to this habit. " Or will you listen," he asks, " to the vain babble of their philosophical authorities, of whom some say that fire is God ? They mistake their own destination for the deity." Again, one Hermias who published a *Mockery of the Pagan Philosophers*, selects Pythagorean mathematics among other things as the object of his attack. " Does Pythagoras measure the universe ? I feel the enthusiasm seize me too. I care no more for home and fatherland, for wife and children. I myself mount up into the sky, borrow the rod of Pythagoras, and make a beginning by measuring fire. That Zeus has measured it is not enough. If I, great creature that I am, great body that I am, great spirit that I am, do not mount up to heaven and measure the sky, all is over with the providence of God." In this way was the presumption of the natural philosophers rebuked.

In the work *On Christian Doctrine* (A.D. 427) of St. Augustine the vexed question of pagan knowledge versus Christian received a solution that was destined to fix for more than a thousand years the

mental framework of the new Europe. The
necessity for St. Augustine is not to know the world
but to know God, and to know God it is necessary,
not to study nature, but Holy Writ. All other
knowledge is to be subordinated to this end. Greek
and Hebrew must be studied, but only because
these are the languages of Holy Writ. Greek pagan
literature, with all its tradition of scientific knowledge, ceases to have any claim on the attention of
Christian men. Nature too must be studied, in
so far as it is useful for understanding the Scriptures.
Examples Augustine himself gives are that a knowledge of serpents helps us to understand *Matthew*
x, 16 (Be ye wise as serpents and harmless as doves)
and of hyssop Psalm l, 9 (Purge me with hyssop
and I shall be clean). So much for Aristotle and
Dioscorides. The mechanical arts require no
special study. The acquaintance we pick up with
them in the ordinary course of life will enable us
to understand the allusions to them in the Scriptures. In general it is not unfair to say that he
looks upon nature simply as a possible aid to the
understanding of the Bible. From his point of
view this attitude is easily intelligible. The world
for him was but a temporary stage on which is
played out the drama of man's eternal destiny.
The libretto was the Bible ; the stage was at any
moment due to disappear. This is the meaning of
his fundamental principle that " immutable wisdom

is to be preferred to mutable." He concludes with a famous judgment which supersedes all pagan learning as superfluous. "Whatever knowledge man has acquired outside Holy Writ, if it be harmful it is there condemned ; if it be wholesome, it is there contained."

This view was still in favour with Milton at the end of the seventeenth century. In *Paradise Regained*, Book IV, he makes the Tempter suggest to the Son of God :

> All knowledge is not couch't in Moses Law,
> The Pentateuch or what the Prophets wrote ;
> The Gentiles also know, and write, and teach
> To admiration, led by Nature's light . . .
>
> To whom our Saviour sagely thus replied.
> Think not but what I know these things, or think
> I know them not ; not therefore am I short
> Of knowing what I ought : he who receives
> Light from above, from the fountain of light,
> No other doctrine needs, though granted true ;
> But these are false, or little else but dreams,
> Conjectures, fancies, built on nothing firm.

Christianity thus jettisoned the whole cargo of pagan science which we in modern times labour so assiduously to recover. Strange as this reaction of Christianity seems, it does not seem impossible that humanity should repeat its mistake, so little does it succeed in using the power science gives to promote its own happiness. Nevertheless one or two conclusions emerge from the story of the creation, loss, and re-creation of science within the few thousand years that comprise the record of civilization. The first is that the activity of the scientist is rarely now in itself regarded as an evil or indifferent thing. On the contrary a knowledge of nature is more and more looked upon as an indispensable element in sane living and in sane thinking. Two thousand years ago Lucretius said of the old Ionian philosopher-scientists that " out of their hearts' holy of holies they had given answers with more sanctity and truth than the priestess of Apollo in her shrine at Delphi." And it may be said that the moral authority of the scientist is now far higher in Europe than that of the priest. The severity of the discipline imposed by scientific research has even begun to receive acknowledgment from its traditional enemy, the Church. Baron von Hügel, a generation ago, in a great crisis in religious thought in which the absence in the orthodox party of the scientific conscience seemed plain to many, advanced the remarkable view that " the study of an experimental science, or critical scholarship, ought to have its normal necessary place in the very theory of spirituality."—" Nature, history, all subjects of research first of all, present us with laws, with things, and neither the clamours of the petty self in front of them, nor, at first sight, the intimations of the Divine Person behind and above them, find here an echo or a place. Nothing breaks the purifying power of the thing, and its apparent fatefulness ; the apparent determinism of the phenomena and the mentally and emotionally costing character of their investigation."

The second point is that it is recognized, in Dr. Sarton's words, that " the acquisition and systematization of positive knowledge is the only human activity which is truly cumulative and progressive." This too carries with it its inevitable moral consequences. Just as all scientific investigation is fruitless which is not pursued in a spirit of truth, so the results of all scientific endeavour are wasted if the continuity of tradition cannot be assured. It is of the very essence of science to be a co-operation and that not only of the men of the same generation, but of the generations successively.

The quest of positive knowledge, then, imposes on those who are susceptible to its influence, its own humanism. It is found, as we said in an early chapter, to be the product of a certain type of man in a certain type of society. It frustrates itself in so far as it is not scrupulously truthful to objective fact, and in so far as it is not a collaboration that unites contemporaries in all lands and the successive generations in all time. It is essentially social, since its appeal is to the common reason in all men, and it repudiates all claim to validity except such as the collective experience of mankind confirms in practice. It failed in antiquity because it was the narrowly-conceived occupation of a privileged class which did not take root in the general education nor yield the benefits it is capable of to men at large. Should it fail again it will be for the same reason. But it seems impossible that a stable society should ever establish itself without the positive knowledge science has to offer as a universal subject of education and the morality science imposes as an element in its law.

SCHEDULE OF DATES

A. *Ancient Cultures of the Near East.*

4236 B.C. Introduction in Egypt of the thirty-day month and the first practical calendar.

4000–3500 B.C. Sumerians reclaim the marshes about the mouths of the Tigris and Euphrates.

3500 B.C. Alphabet of twenty-four letters introduced in Egypt. Sumerian picture-writing, passing soon into phonetic cuneiform signs.

3360 B.C. Copper mines of Sinai begin to be exploited by the Egyptian kings.

3000-2500 B.C. Age of the building of the great pyramids.

2000 B.C. Feudal Age in Egypt. To this age belong the surgical and mathematical papyrus rolls. The Babylonian law-giver Hammurabi is dated 1948–1905 B.C.

B. *Greek Culture : Literature that preceded or accompanied the rise of Greek Science.*

Epic. The *Iliad*, *Odyssey* and *Theogony* were all in existence before 700 B.C.

Lyric. Archilochus, floruit 648 B.C.
Sappho and Alcæus, fl. 600 B.C.

Attic Tragedy. The lives of Æschylus, Sophocles, and Euripides are comprised within the years 525–406 B.C.

Historiography. Herodotus c. 484–425 B.C.
Thucydides 471 401 B.C.

C. *Principal names in the history of Greek science, together with others mentioned in the text.*

Thales, fl. 585 B.C.
Anaximander, c. 610–546 B.C.

SCHEDULE OF DATES

Anaximenes, fl. c. 546 B.C.
Pythagoras, c. 572–500 B.C.
Xenophanes, c. 580–480 B.C.
Heraclitus, fl. c. 504 B.C.
Parmenides, fl. c. 504 B.C.
Anaxagoras, 500–428 B.C.
Zeno the Eleatic, fl. c. 464 B.C.
Protagoras, 480–411 B.C.
Gorgias, c. 483–375 B.C.
Empedocles, fl. c. 445 B.C.
Democritus, fl. c. 420 B.C.
Hippocrates of Chios, fl. in Athens c. 450–430 B.C.
Hippocrates of Cos, c. 460–380 B.C.
Socrates, 469–399 B.C.
Archytas, fl. c. 380 B.C.
Plato, 427–347 B.C.
Eudoxus of Cnidos, c. 408–355 B.C.
Heraclides of Pontus, c. 388–312 B.C.
Aristotle, 384–322 B.C.
Theophrastus, succeeded Aristotle 322 B.C.
Epicurus, 341–270 B.C.
Strato, succeeded Theophrastus, 288 B.C.
Pytheas of Marseilles, fl. 330 B.C.
Euclid, fl. between 323 and 285 B.C.
Herophilus, fl. between 323 and 285 B.C.
Erasistratus, somewhat later than Herophilus.
Archimedes, 287–212 B.C.
Seleucus the Babylonian astronomer, fl. c. 250 B.C.
Eratosthenes, c. 273–192 B.C.
Apollonius, fl. c. 220 B.C.
Dionysius Thrax, fl. c. 130 B.C.
Hipparchus, fl. 125 B.C.
Marcus Terentius Varro, 116–27 B.C.
Lucretius, 98–55 B.C.
Cicero, 106–43 B.C.
Celsus, fl. c. A.D. 30.
Pomponius Mela, fl. A.D. 43.
Seneca, died A.D. 65.

SCHEDULE OF DATES

Pliny the elder, A.D. 23–79.
Dioscorides, fl. A.D. 50.
Plutarch, died A.D. 125.
Ptolemy, died after A.D. 161.
Galen, A.D. 129–199.
Bardesanes, A.D. 154–222.
Origen, c. A.D. 185–254.
St. Augustin, A.D. 354–430.
Q. Julius Hilarianus, fl. c. A.D. 397.
Cosmas Indicopleustes, fl. c. A.D. 540.

BIBLIOGRAPHY

J. H. BREASTED : *Ancient Times*, Ginn and Co., Boston, 2nd ed., 1935.

GEORGE SARTON : *Introduction to the History of Science*, Vol. I, The Williams and Wilkins Co., Baltimore, 1927.

ABEL REY : *Science Orientale*, La Renaissance du Livre, Paris, 1930.

—— *La Jeunesse de la Science grecque*, Paris, 1935.

—— *La Maturité de la Pensée scientifique en Grèce*, Paris, 1939.

C. M. BOWRA : *Ancient Greek Literature*, The Home University Library, 1933.

W. A. HEIDEL : *The Heroic Age of Science*, The Williams and Wilkins Co., Baltimore, 1933.

ARNOLD REYMOND : *The Sciences in Græco-Roman Antiquity*, Methuen & Co., 1927.

HERMANN DIELS : *Antike Technik*, Leipzig, 2nd ed., 1920.

ALBERT NEUBURGER : *The Technical Arts and Sciences of the Ancients*, Methuen & Co., 1930.

J. L. HEIBERG : *Mathematics and Physical Science in Classical Antiquity*, Oxford University Press, 1922.

T. L. HEATH : *Manual of Greek Mathematics*, Oxford, 1931.

—— *Greek Astronomy*, Dent and Sons, London 1932.

—— *Aristarchus of Samos*, Clarendon Press, Oxford, 1913.

CHARLES SINGER : *The Evolution of Anatomy*, Kegan Paul, Trench, Trubner and Co., 1925.

—— *Short History of Medicine*, Clarendon Press, Oxford, 1928.

250

BIBLIOGRAPHY

CHARLES SINGER: *Greek Biology and Greek Medicine*, Oxford University Press, 1922.

A. J. BROCK: *Greek Medicine*, Dent and Sons, London, 1929.

CARY and WARMINGTON: *Ancient Explorers*, Methuen & Co., 1929.

J. M. STILLMAN: *The Story of Early Chemistry*, D. Appleton and Co., New York and London, 1924.

J. L. E. DREYER: *History of the Planetary Systems from Thales to Kepler*, Cambridge, 1906.

A. E. TAYLOR: *Plato, the Man and His Work*, Methuen and Co., 1926.

WERNER JAEGER: *Aristotle*, Clarendon Press, Oxford, 1934.

W. D. ROSS: *Aristotle*, Methuen and Co., 2nd ed., 1930.

JOHN BURNET: *Early Greek Philosophy*, Black, 4th ed., 1945.

BRUNET ET MIELI: *Histoire des Sciences. Antiquité*, Payot, Paris, 1935.

IVOR THOMAS: *Greek Mathematical Works*, Loeb Library, vol. I, 'Thales to Euclid', 1939; vol. II, 'Aristarchus to Pappus of Alexandria', 1941.

F. GRANGER: *Vitruvius* (2 vols.), Loeb Library, 1931 and 1934.

INDEX

252

INDEX

INDEX

254

INDEX

INDEX

Printed in Great Britain by Butler & Tanner Ltd., Frome and London